The Sun is 'dying', build a New World

Science Exploration by Rolf A. F. Witzsche

The Sun is 'dying'
build a New World

This book contains the transcript text with all images included, of the Rolf A. F. Witzsche exploration video production with the above title:

see: http://www.ice-age-ahead-iaa.ca/

Lead in:

Climate Change is measured reality, but it isn't manmade. It is caused by changing solar activity, which is now increasingly collapsing, and the climate is collapsing with it. We are now in the boundary zone of the process, of the weakening Sun, 'dying' towards the next Ice Age. The zone of roughly 50 years began when the last solar global warming ended in the 1990s, and solar activity sharply reversed. We are currently half-way through Phase 1 of the reversal, which is unfolding like a free-fall collapse.

The Global Warming died in 1998.
It died with the Sun beginning to die back
towards its Ice Age phase shift in the 2050s.
The dying of the Sun is real -
Prepare for it now!

We know with great certainty that all climate changes, with a few minor exceptions, were and still are caused by the Sun - almost exclusively by changing solar cosmic-ray flux affecting cloud forming on Earth - because we have measured the cause for the process in numerous ways. We have 'measured' the Sun's activity in sunspot numbers, and in historic ratios of radio-isotopes that are exclusively generated by solar cosmic-ray flux in the Earth's atmosphere. We have also 'measured' the Sun in real-time with satellites in space, with radio-telescopes on the ground, and as of late by measuring the neutron-density in the atmosphere as a real-time proxy for changing solar cosmic-ray flux that releases neutrons in the atmosphere by interaction. And what do all of these measurements tell us? They tell us that the Sun is diminishing fast, and that the climate on Earth is correspondingly getting colder, year after year, and more so from 2018 onward, for potentially another 30 years till the Ice Age begins, and things get really bad.

We need to build a New World with new agricultures,
as the current agricultures are beginning to fail evermore,
worldwide, under the fast weakening Sun.

All this means that our agriculture is in danger, and that when food production grinds to a halt, especially in the climate vulnerable regions, entire nations potentially cease to exist with no place for the populations to migrate to, unless a large-scale Plan-B creates a New World for them, with new technological infrastructures for new agricultures that the changing climate cannot affect; and with new cities, and so forth, for continued human living.

Plan-B is critical, because the current Plan-A is to do nothing, with the song, "let the people die."

Plan-B, however is blocked by the Manmade Global Warming hoax that is projected from the highest levels of the United Nations. While over 50,000 scientists have declared their opposition to the Manmade Global Warming doctrine in numerous petition projects and protest statements, the opposition movements do actively support the core assumption of the Manmade Global Warming doctrine, which is the theory of the invariable Sun.

For Plan-B to have a chance, the fundamental blocking factor that is purely theoretical without exclusive evidence to stand on, and which tends to be politically inspired, needs to be overcome.

The needed change may not come from the top through the United Nations, however. It may well be that a Plan-B option may be implemented one day soon, unilaterally, by one of the most advanced nations on Earth simply stating the project, that the rest of the world would than gladly join up.

Major Topics

1- The Variable Sun: Our Climate Forcer

2 - Measuring Historic Solar Activity in Proxy

3 - Changing Solar Activity, Measured in Real-Time, in Proxy

4 - Climate Change: Forced by Changing Solar Cosmic-Ray Flux - almost exclusively

5 - Real-Time Experienced Climate Consequences

6 - Climate Collapse - Agricultural Collapse

7 - Exit from the Boundary Zone - Plan-B

8 - Deadly Blocking Factor: The Global Warming Hoax

9 - Plan-B Unfolding: Spontaneous, Unbounded, Cultural and Industrial Revolution

Contents

Ice Age Boundary Zone

Between the end of solar global warming and the start of the near Ice Age unfolds the Ice Age Boundary Zone.

Yes, there was such a thing as Global Warming

Yes, there was such a thing as Global Warming. But is wasn't manmade. It was caused by the Sun.

Part 1 - The Variable Sun: Our Climate Forcer

Part 1

The Variable Sun

Our Climate Forcer

Part 1 - The Variable Sun: Our Climate Forcer

The Sun's Global Warming ended in the late 1990s

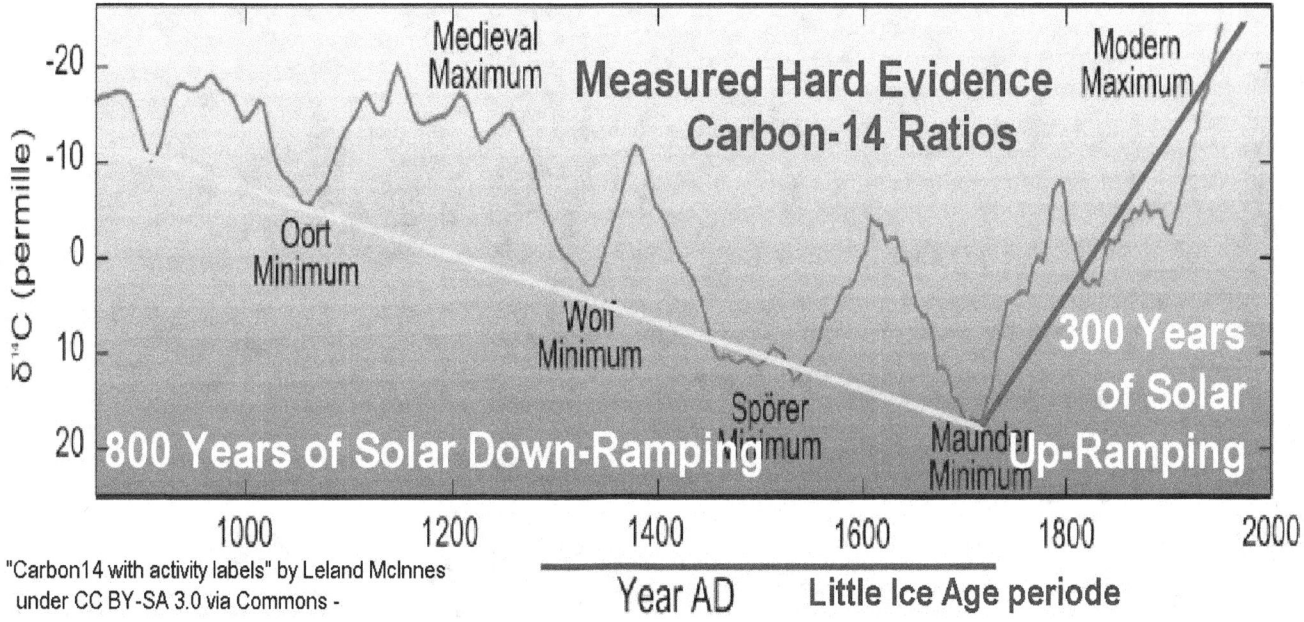

Changing solar cosmic-ray flux, measured in carbon-14 ratios, shows direct inverse relationship with known cold-climate events

"Carbon14 with activity labels" by Leland McInnes under CC BY-SA 3.0 via Commons -

The Sun's Global Warming broke the Little Ice Age and lasted for nearly 300 years. It ended in the late 1990s.

Grand global warming by the Sun, is history

Wheat harvest on the Palouse, Idaho, USA

Food for eating!

wikipedia

This grand global warming by the Sun, is history now. Remember it fondly! It won't be happening again in your lifetime! The Earth is getting colder every year. In the course of it, agriculture becomes endangered.

We are now in the boundary zone

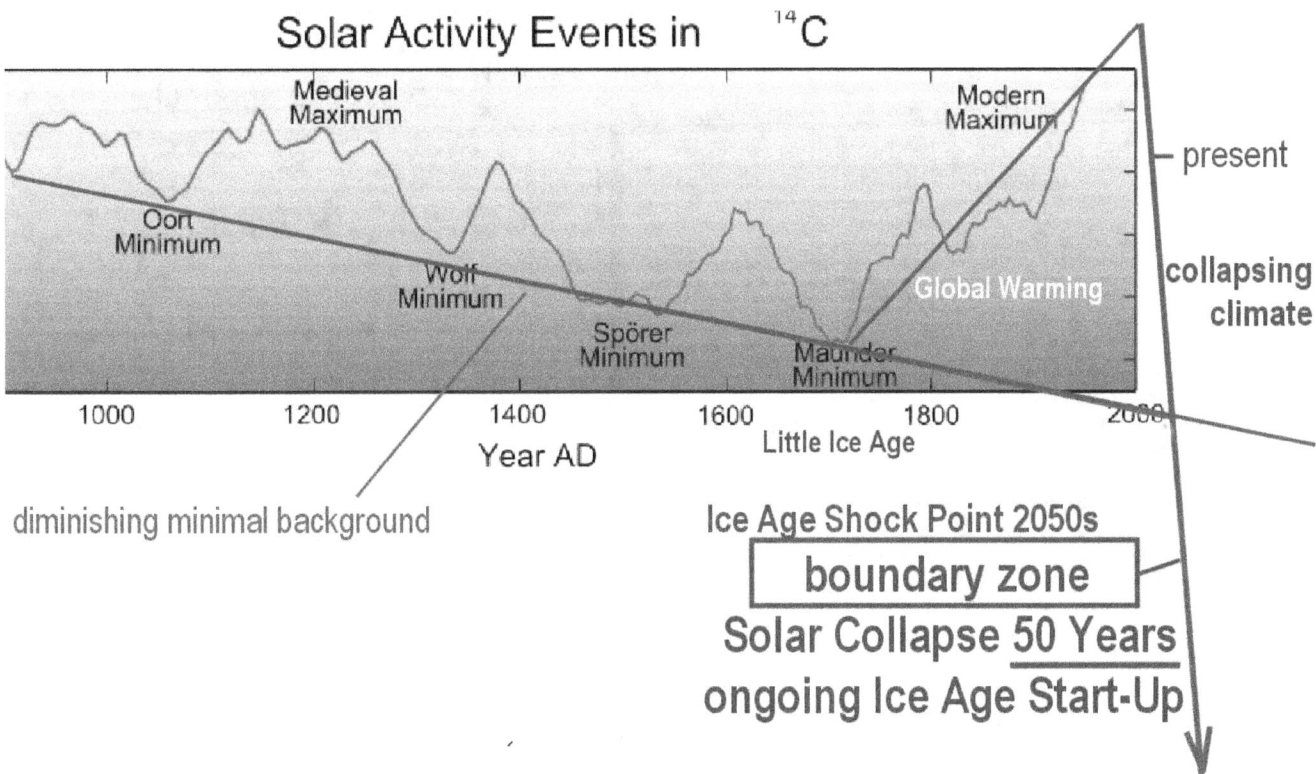

We are now in the boundary zone that marks the interval in time between the end of the last solar global warming and the start of the next Ice Age, in potentially the 2050s.

Boundary zone is not the transition zone

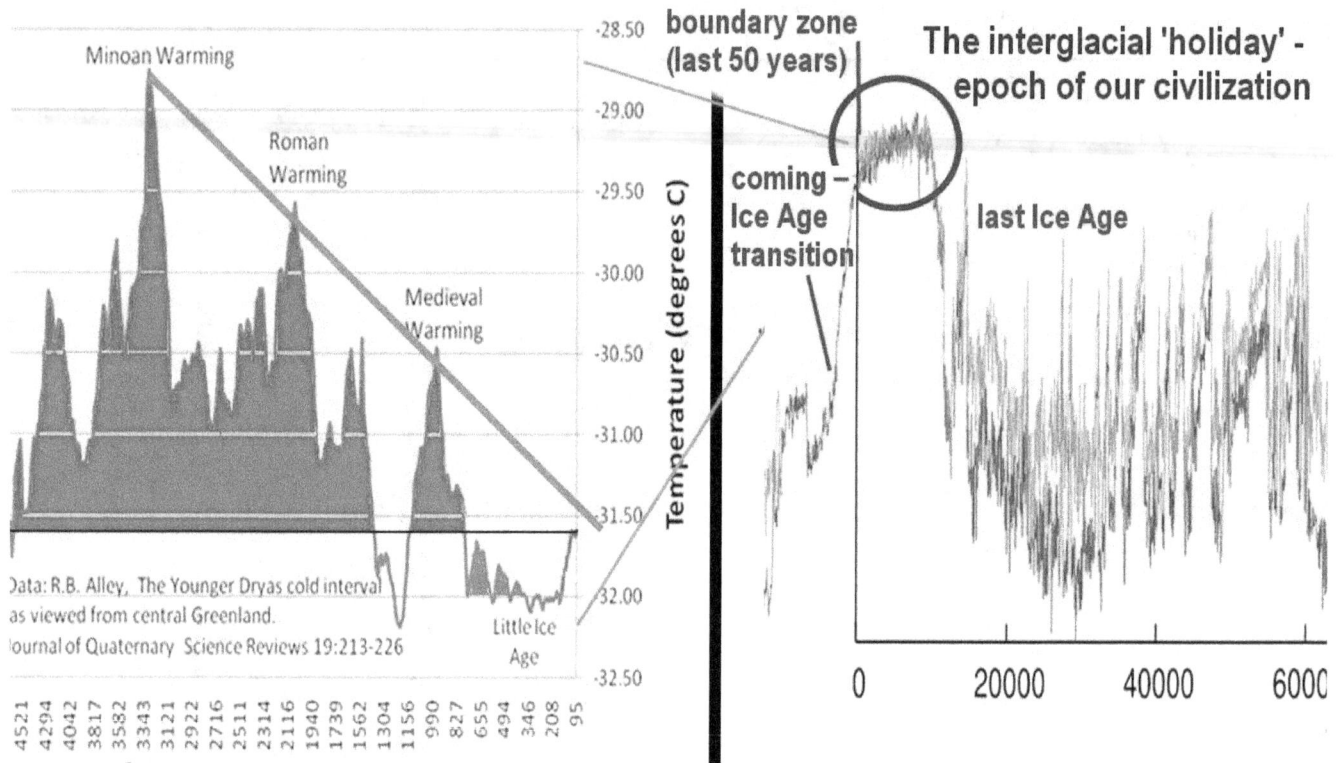

Last Ice Age

The coming Ice Age transition

The boundary zone (the last 50 years)

This boundary zone is not the transition zone in which the next Ice Age unfolds. The Ice Age transition is an enormous event that takes us into a completely different world that no one can likely imagine.

The boundary zone, in comparison, comprises the timeframe of the last 5 decades of the present interglacial climate, before the Ice Age transition begins. This is the timeframe in which the dynamic system that powers the Sun, begins to weaken irreversibly.

The visible results are fringe effects that appear faintly at first, but then happen evermore rapidly, and with increasing severity, as the system that powers the Sun diminishes evermore towards its final phase shift by which the Earth becomes radically transformed.

Before that phase shift happens, the weakening of the system that powers the Sun becomes expressed in numerous escalating climate effects that are beginning to affect our agricultures and with it our food supplies. That's the timeframe of the boundary zone.

In the boundary zone our Sun is 'dying'.

To put it more dramatically, in the boundary zone our Sun is 'dying'. That's what we are facing now. There is nothing we can do to halt or reverse the process that is diminishing our Sun. However, we can build us a New World with technological infrastructures placed across the Equatorial seas that enables our continued living and unending development under a lesser Sun.

In speaking of solar global warming

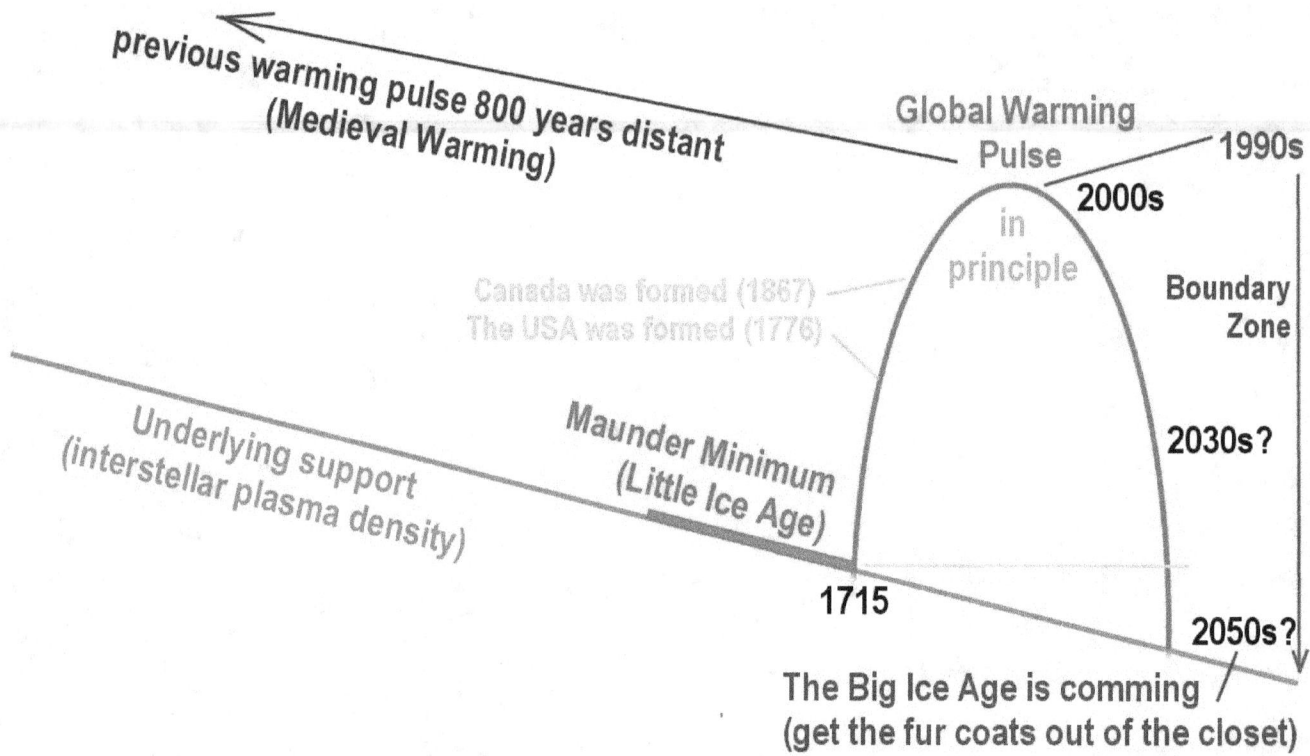

In speaking of solar global warming, I am referring to the Great Global Warming of the Earth that had saved humanity from a potential Ice Age transition in the 1700s, which had terminated the little Ice Age.

In the 1700s the Sun became strengthened by a plasma resonance effect within the system that powers the Sun. When the resonance effect ended, the process began to unfold in reverse. In this reversal, the Sun is diminishing back towards the level of the Little Ice Age. And since there is little background support left in the system, the Sun will diminish beyond the Little Ice Age into to the Big Ice Age that we had just barely missed in the 1700s.

We are in that kind of boundary zone.

We are 20 years into the Ice Age boundary zone

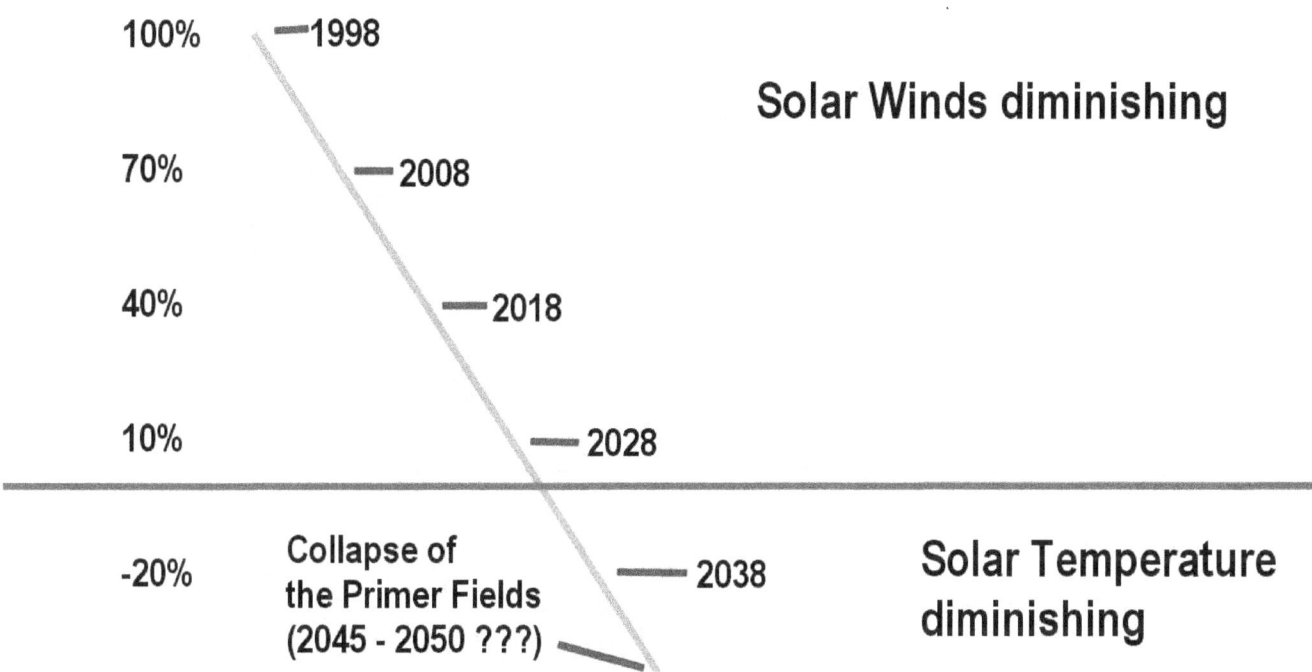

We are presently 20 years into the Ice Age boundary zone. As I said before, the boundary zone began when the last period of solar global warming ended and solar activity began to reverse. The reversal became evident in the collapsing solar wind pressure at the rate of 30% per decade, that had been measured in space in real time by the Ulysses spacecraft between 1998 and 2008.

Global Warming caused by cosmic factors

Changing solar cosmic-ray flux, measured in carbon-14 ratios, shows direct inverse relationship with known cold-climate events

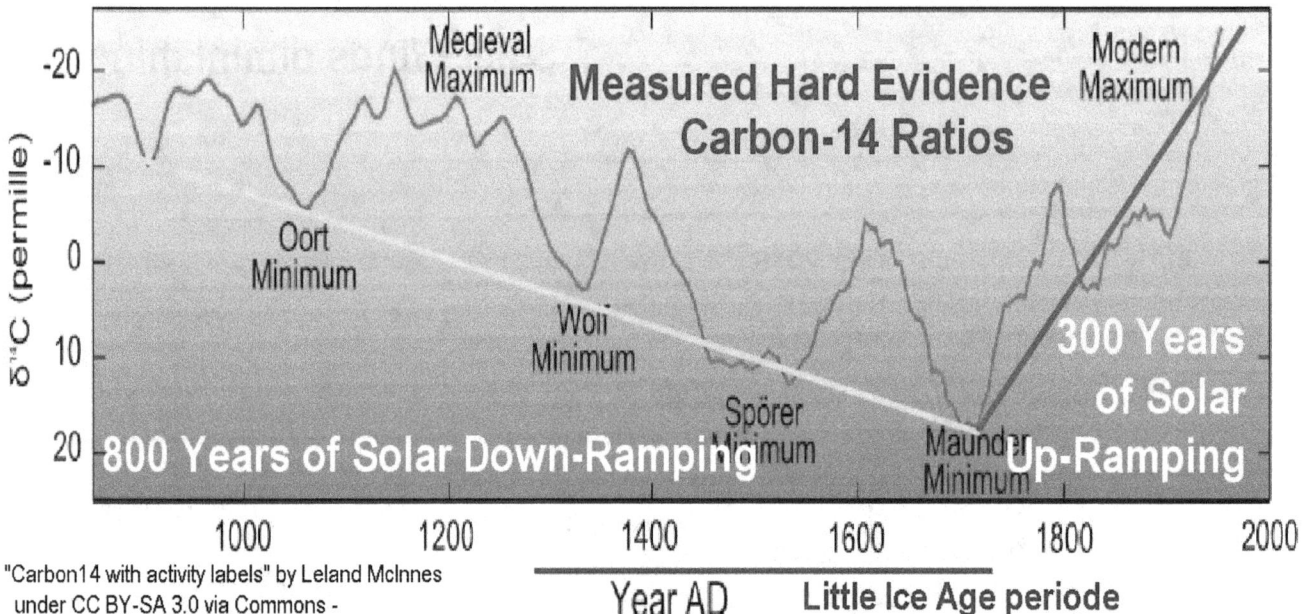

"Carbon14 with activity labels" by Leland McInnes under CC BY-SA 3.0 via Commons -

The Great Global Warming that had saved us from the Big Ice Age, in the 1700s had been caused by cosmic factors that had up-ramped solar activity in a big way. This up-ramping was real. It was powerful. The resulting Solar Global Warming gave us the most productive climate for agriculture in more than a thousand years. That's what had ended in the 1990s.

Far into this last period of global-warming

Changing solar cosmic-ray flux, measured in carbon-14 ratios, shows direct inverse relationship with known cold-climate events

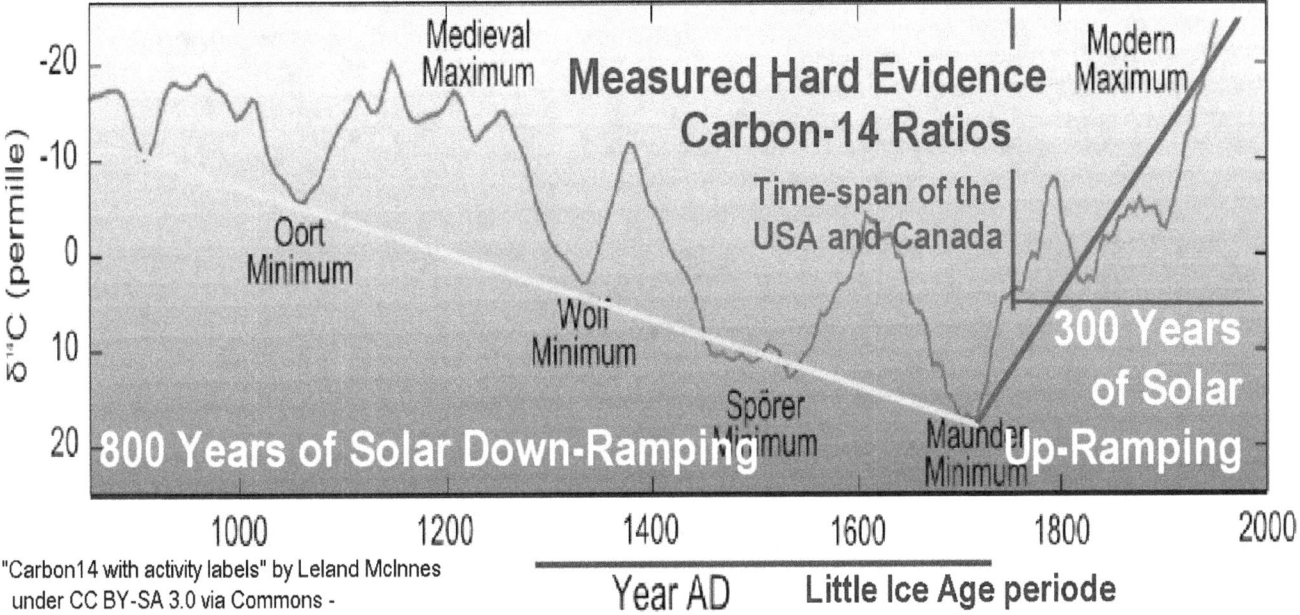

"Carbon14 with activity labels" by Leland McInnes under CC BY-SA 3.0 via Commons -

Far into this last period of global-warming, the USA became created. And soon thereafter the federation of Canada was enacted. At the same time Europe developed its second renaissance that may be termed the Westphalian Renaissance. This means that the climate foundation on which these nations were built up on, is now fading, which endangers the nations.

The Cosmic Global Warming is now irreversibly ending

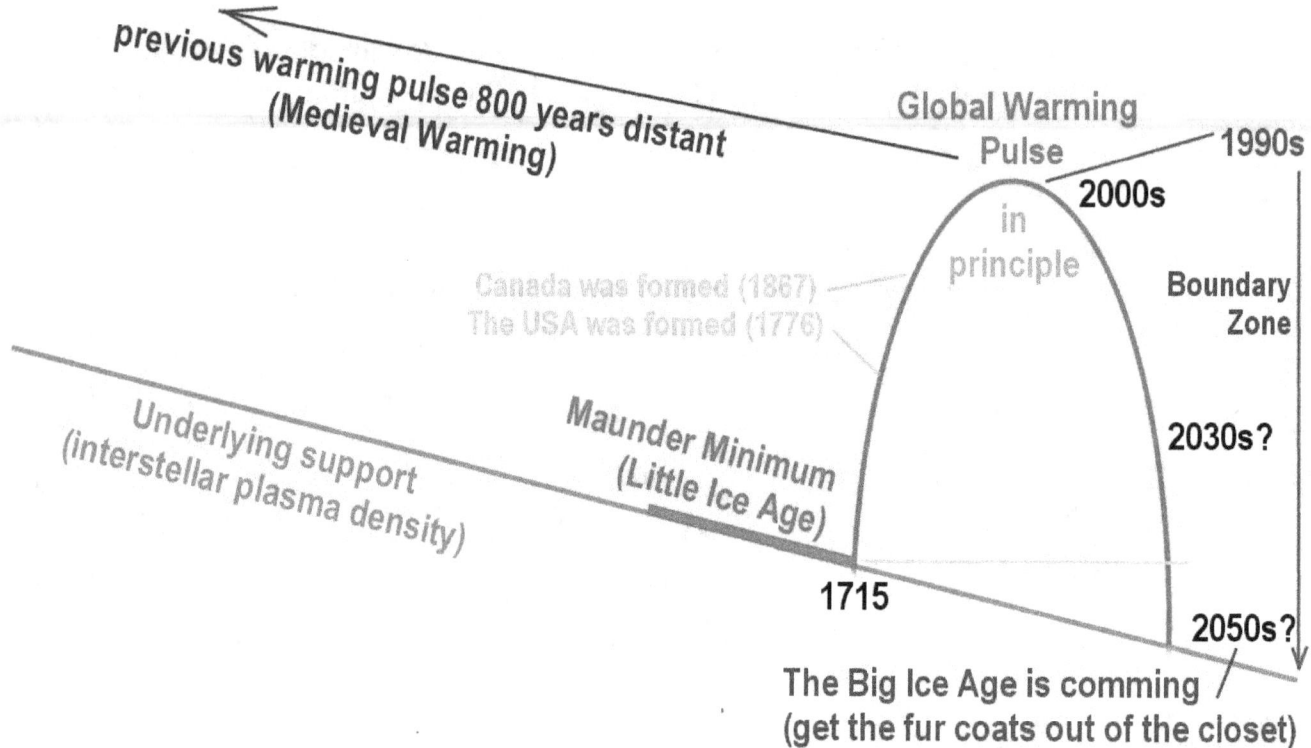

Canada was formed (1867)

The USA was formed (1776)

Boundary Zone 1990s to present

Unfortunately, the Cosmic Global Warming period that was brought about by an effect within the solar system - a type of global warming pulse, within which the USA and Canada were formed and had developed - is now irreversibly ending. We simply cannot get it back. The system that powers our Sun is now rapidly diminishing, as the global warming pulse is drawing to its end.

Interglacial climate is diminishing evermore rapidly

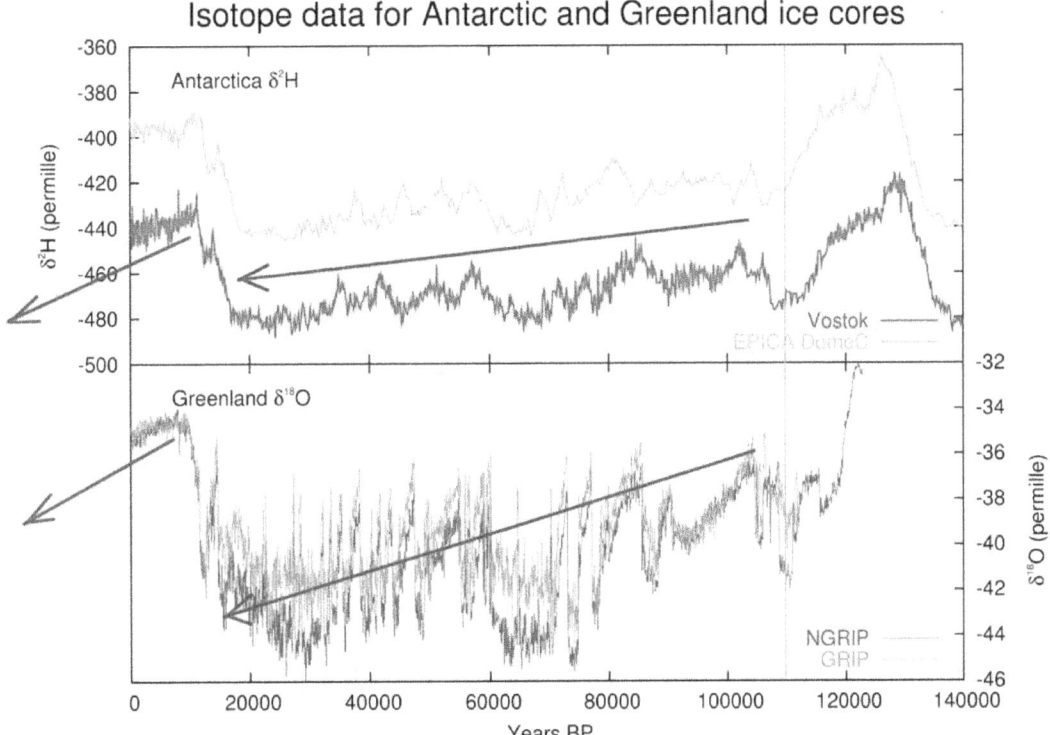

With the interglacial climate having been gradually diminishing ever since the interglacial optimum 8,000 years ago, we find in ice core samples that the interglacial climate is diminishing evermore rapidly now as we are coming to the end of it. This means that the underlying support for the solar system no longer exists for a significant reversal to happen, from the increasing weakening of the Sun.

We, humanity, certainly cannot affect this huge cosmic process by any means that we possess.

So it is that from the 1990s on, we've begun to live in a boundary time-zone towards the inevitable phase shift to the next Ice Age that we cannot escape, but which we have the power to avoid the consequences of, - by us building us a New World with technological infrastructures that are not affected by the climate change to the next Ice Age, - which will be bigger than any ever experienced in recorded history.

Facing the consequences

Facing the consequences

Facing the consequences

Humanity has trapped itself into small-minded concepts

About the image:

The Milankovitch Cycles theory of the Ice Age phenomenon was the first major historic climate theory postulated in the 1920s on the erroneous assumption that the Sun is an invariable constant for all climate considerations. The theory predates the Manmade Global Warming theory by half a century, which is based on the same false assumption, as are many other similar theories. In this sense the Milankovitch Cycles theory represents them all, which are but regurgitations of the same basic false assumption.

In searching for an Ice Age cause, while denying the real cause, the Milankovitch Cycles theory trapped itself into the assumption, that the Earth's 3 minute orbital variations over long periods ranging from 26,000 to 100,000 years-cycles, are the cause for the Ice Age cycles resulting from hemispheric and seasonal variations of the exposure of the Earth to the Sun, even while it is self-evident that the total energy received from the Sun remains the same in every case, which renders the Milankovitch Cycles theory a paradox in itself.

Paradoxes are encountered when the real cause for a phenomenon is ignored or denied, like the measured variable Sun is ignored that is being expressed in corresponding climate fluctuations on Earth. The Manmade Global Warming theory is the most modern of these types of paradoxes that mistake effect for cause and deny the real cause, for which the classical case is the Milankovitch Cycles theory.

Text from the video:

We now face consequences of cosmic phenomena that most of humanity has been taught to believe will never happen. Society is stuck in this blindness for the lack of understanding of the principles that affect our climate, including the next Ice Age that is generally believed to be still thousands of years distant.

Most of humanity has trapped itself into believing in these types of small-minded concepts. Scientists have latched onto the notion that the Sun is too big to change, so that climate variations on Earth, including the Ice Ages, are deemed to be caused purely by Earth-centered factors, such as by changing orbital variations of the Milankovitch cycles theory, or ocean currents variations, and air currents, volcanic activity, greenhouse gases, and similar factors.

As scientists became trapped into mechanistic perceptions

As scientists became trapped into mechanistic, small-minded perceptions, they have closed their eyes to the one overriding factor that is bigger than everything. This overriding factor is the solar factor. The Sun is the biggest climate factor by far, and almost the only climate factor. It is immensely causative, and it is also immensely variable. That's what we have to face and respond to.

Scientists have been recruited to close their eyes

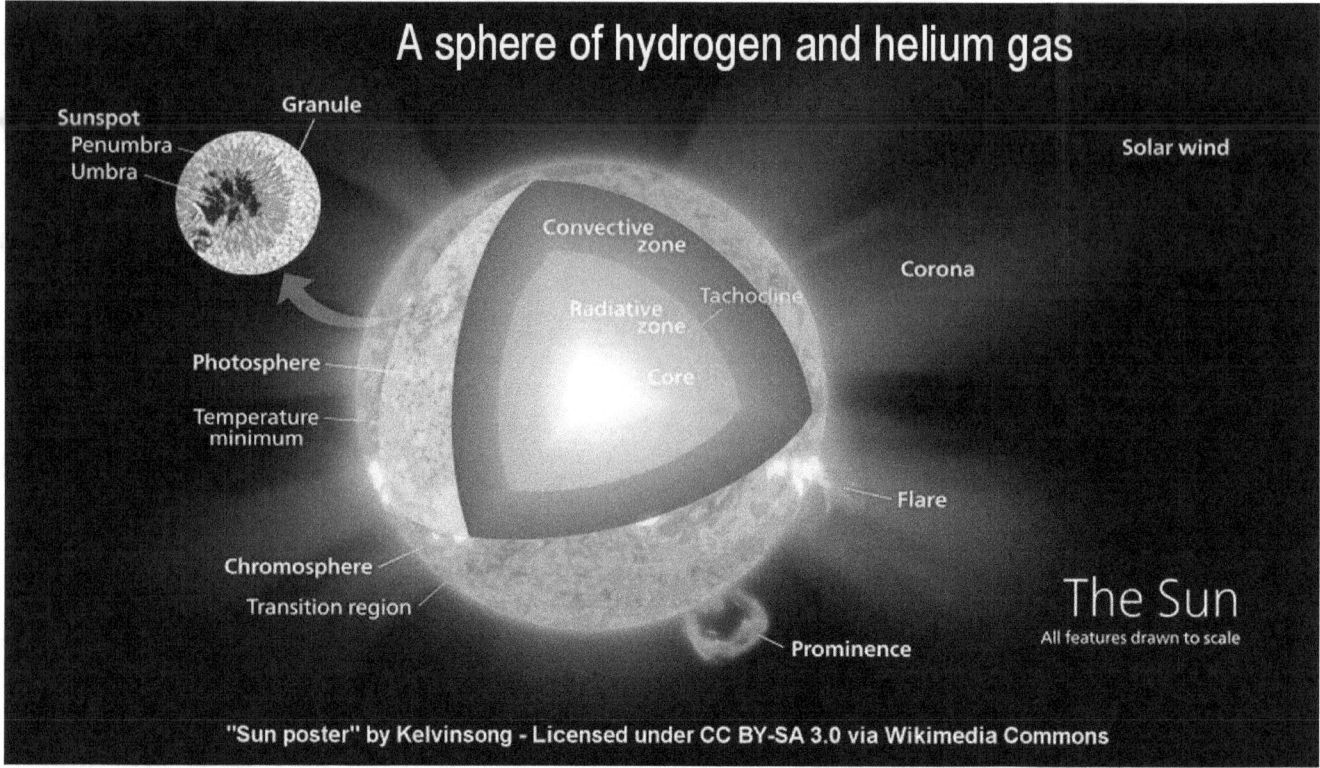

Scientists have been recruited to close their eyes, and to focus not on the Sun as it actually is, but on small things. They have been taught the widely promoted assumption that the Sun is its own master, a giant nuclear fusion furnace that is isolated from the universe and is incapable of fast variation.

In the background of this small thinking, the dogma was developed that the Sun is an invariable constant for all climate considerations, including the Ice Ages. The dogma dictates, that because the Sun does not change, all climate changes without exceptions must necessarily be caused by factors other than the Sun.

As the result, scientists have become a living paradox. They close their eyes to what science has revealed to be true based on physically measured evidence, and believe instead in the opposite.

Evidence of the variable Sun

Evidence of the variable Sun

Evidence of the variable Sun

Measured evidence that the Sun IS changing rapidly

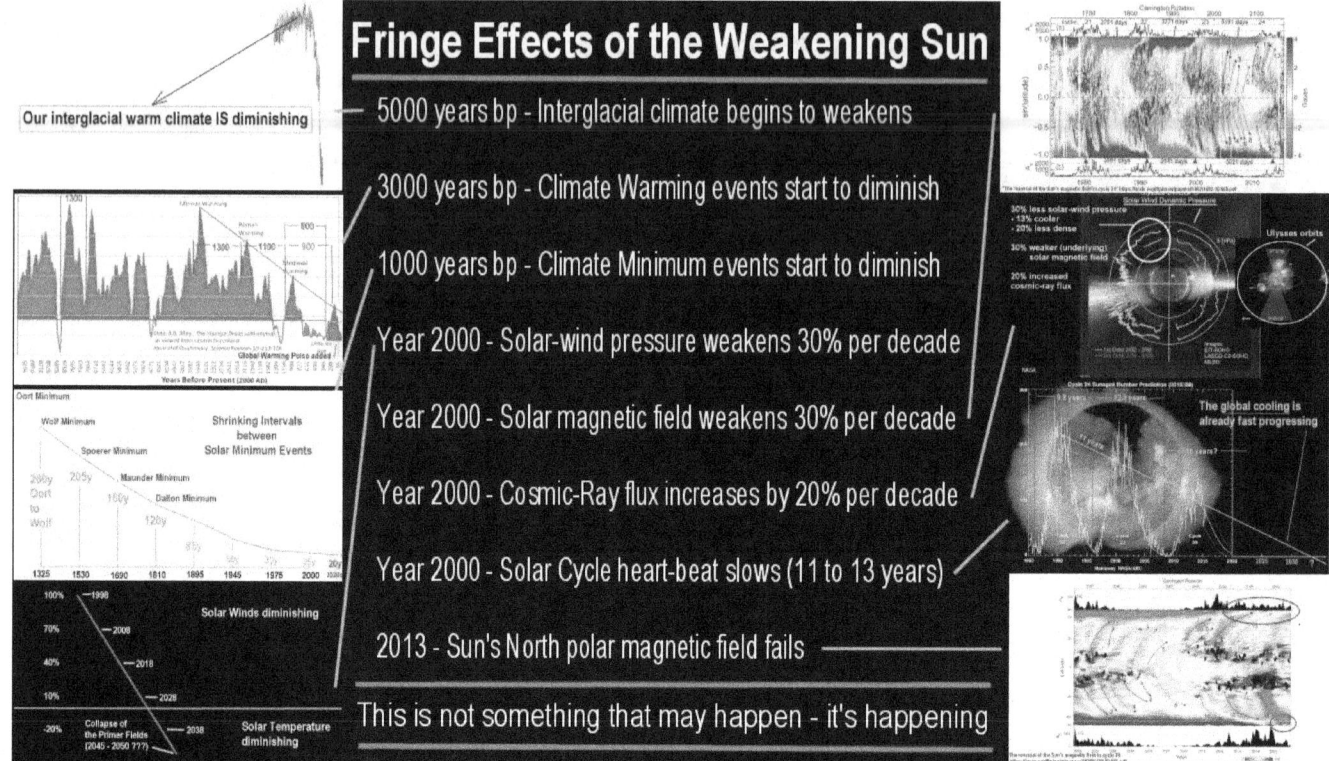

A wide array of physically measured evidence, both modern and historically, has established as an undeniable fact that the Sun is the big and only climate factor on Earth, and that the Sun IS changing rapidly, enormously, and has been changing throughout history in numerous measurable ways.

Let's back-up a bit here

> **Wow! Let's back-up a bit here, and start at the beginning.**

Wow! Let's back-up a bit here, and start at the beginning.

The period of the deep cooling of the Earth

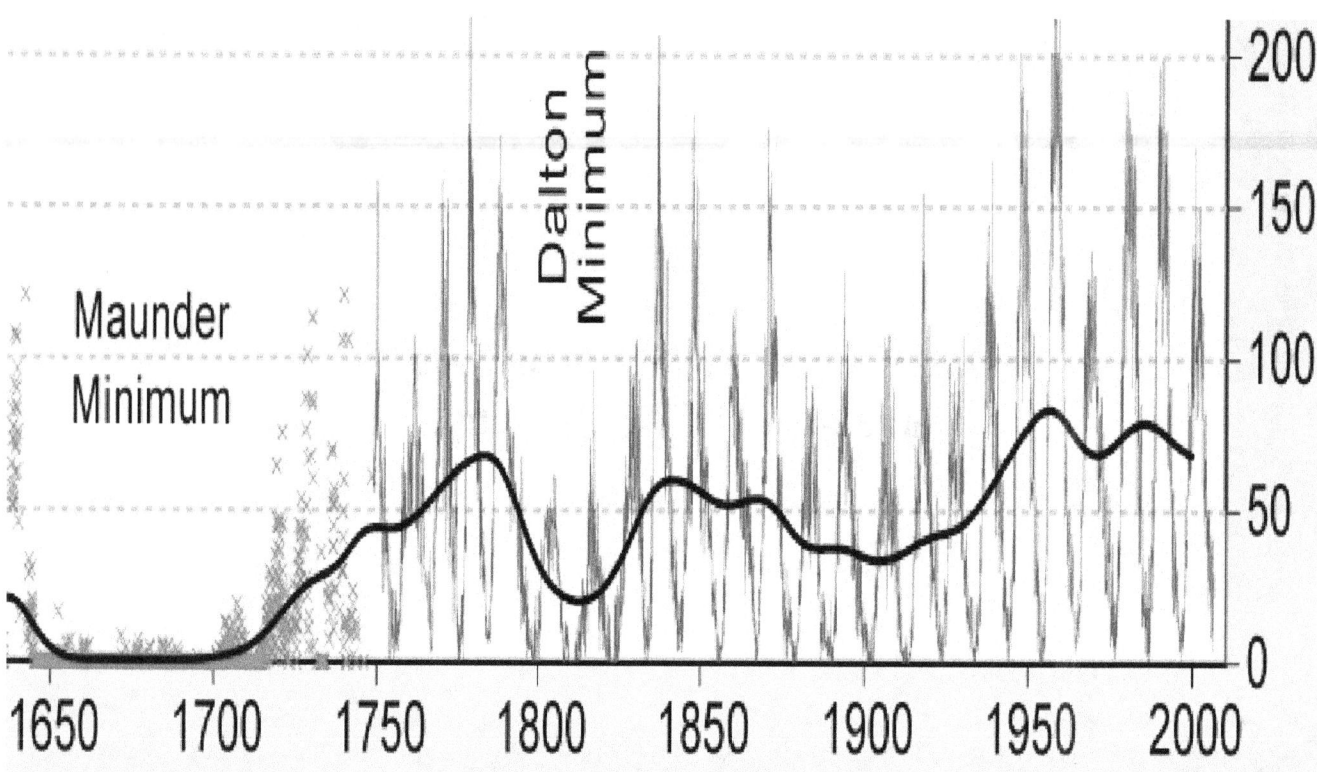

It is a well-established fact that the period of the deep cooling of the Earth in the 1600s, which was the period of the Little Ice Age, had been a period of extremely low solar activity. Sunspot numbers had been carefully recorded at the time. It was noted that for nearly a century almost no sunspots were seen on the surface of the Sun. This period of extremely low, and almost no visible solar activity in terms of sunspots, is referred today as the Maunder Minimum.

Then, suddenly, as is clearly visible in the sunspot numbers counted, the Sun had changed in the 1700s. The Maunder Minimum suddenly ended. In the early 1700s the sunspots were back in big numbers, and with the increasing sunspot numbers the climate on Earth began to warm up dramatically. With the rapid revival of the solar activity, the Little Ice Age was over. With this revival of the Sun, a long period of global warming began.

It has also become possible in recent time to physically measure the intensity of historic solar activity by other means than counting sunspot numbers, and to go farther back in time than the sunspot numbers were recorded.

Part 2 - Measuring Historic Solar Activity in Proxy

Part 2

Measuring Historic Solar Activity

in proxy

Part 2 - Measuring Historic Solar Activity in Proxy

Measuring the changing ratio of two rare radioisotopes

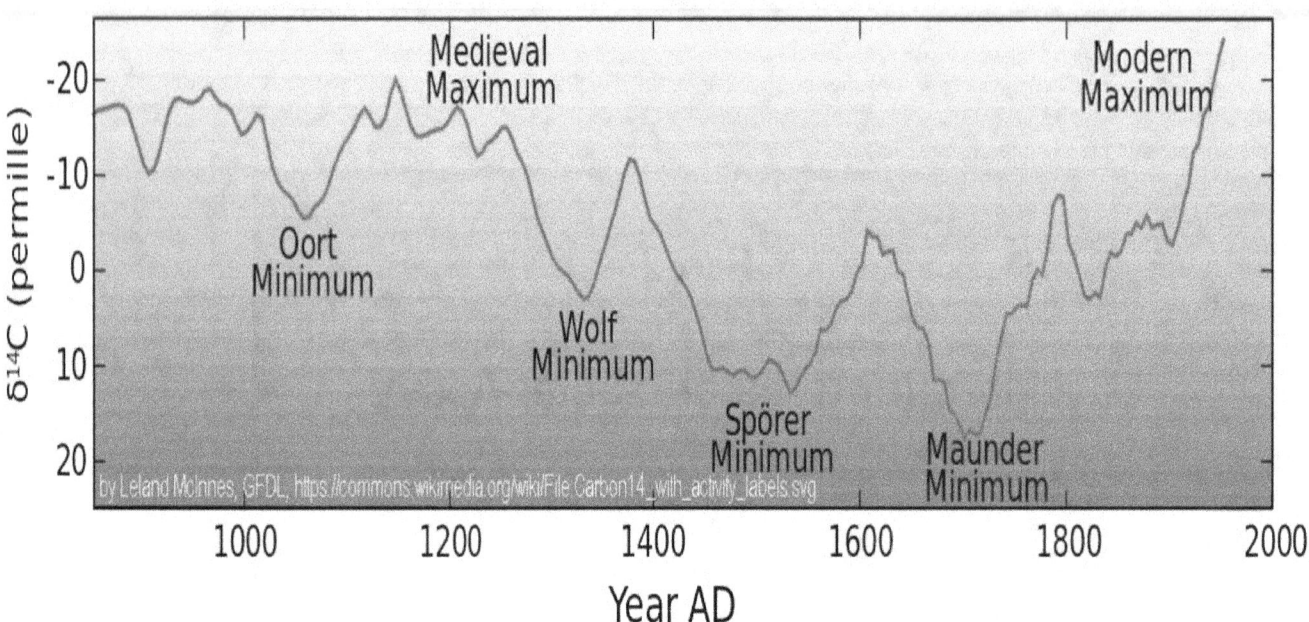

It became possible to measure historic solar activity in proxy, by measuring the changing ratio of two rare radioisotopes, that of carbon-14 and of berillium-10. The density of these isotopes in historic samples, stands as a proxy for the solar activity at the time, because the Sun, and only the Sun, had generated the isotopes in the Earth's atmosphere, which happens when solar cosmic-ray flux interacts with the Earth's atmospheric nitrogen, for example. The density of the Sun-produced isotopes thereby fluctuates with changing solar activity. The fluctuation can be measured.

When our Sun, which is a plasma sun, is weak

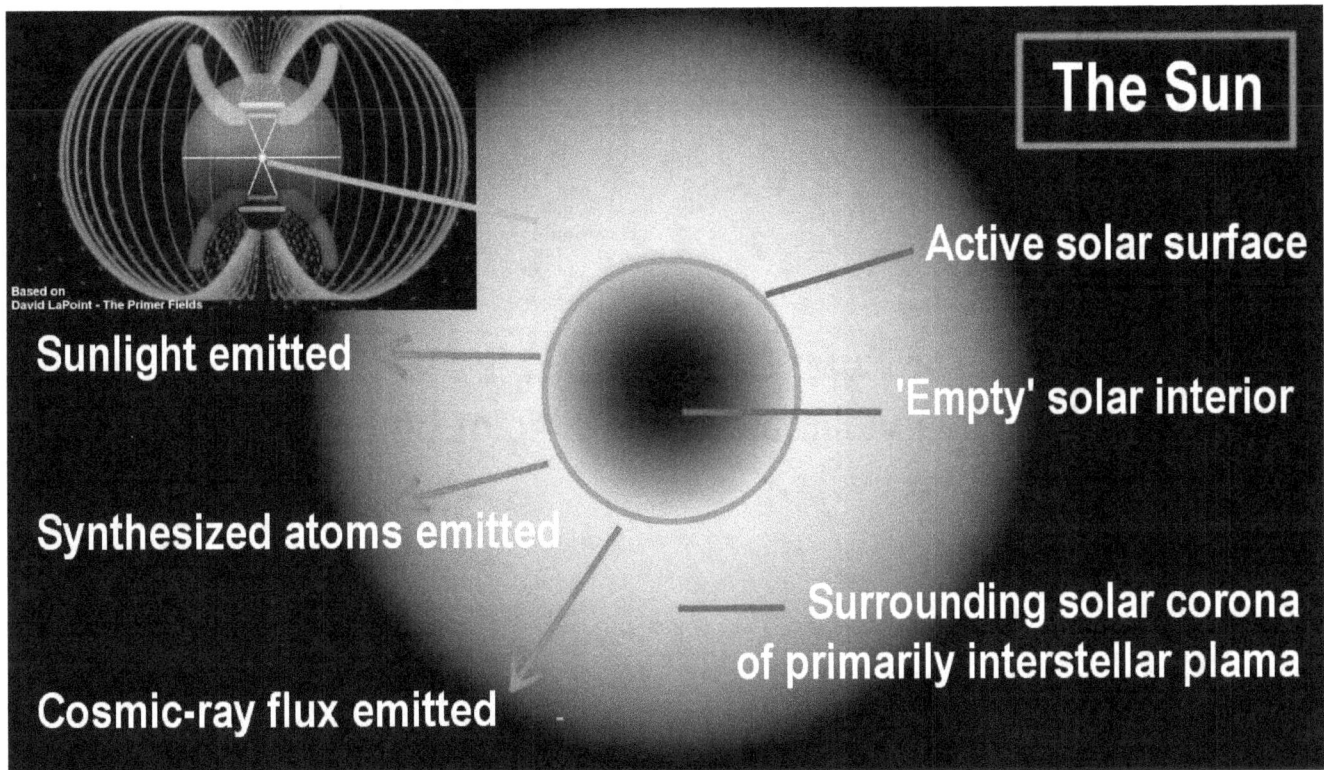

When our Sun, which is a plasma sun, is weak, it has a weak sphere of plasma around it. The plasma shield blocks a portion of the cosmic ray flux that is generated in the Sun's plasma fusions cells on the solar surface. This means that a weak Sun allows a larger volume of solar cosmic-ray flux to penetrate the barrier, which generate larger amounts of the c-14 and be-10 isotopes in the atmosphere on Earth.

The Solar Cosmic-Ray Flux

The Solar Cosmic-Ray Flux IS the primary climate-forcing factor on Earth

The Solar Cosmic-Ray Flux IS the primary climate-forcing factor on Earth

Prove concretely that the Sun is a highly variable

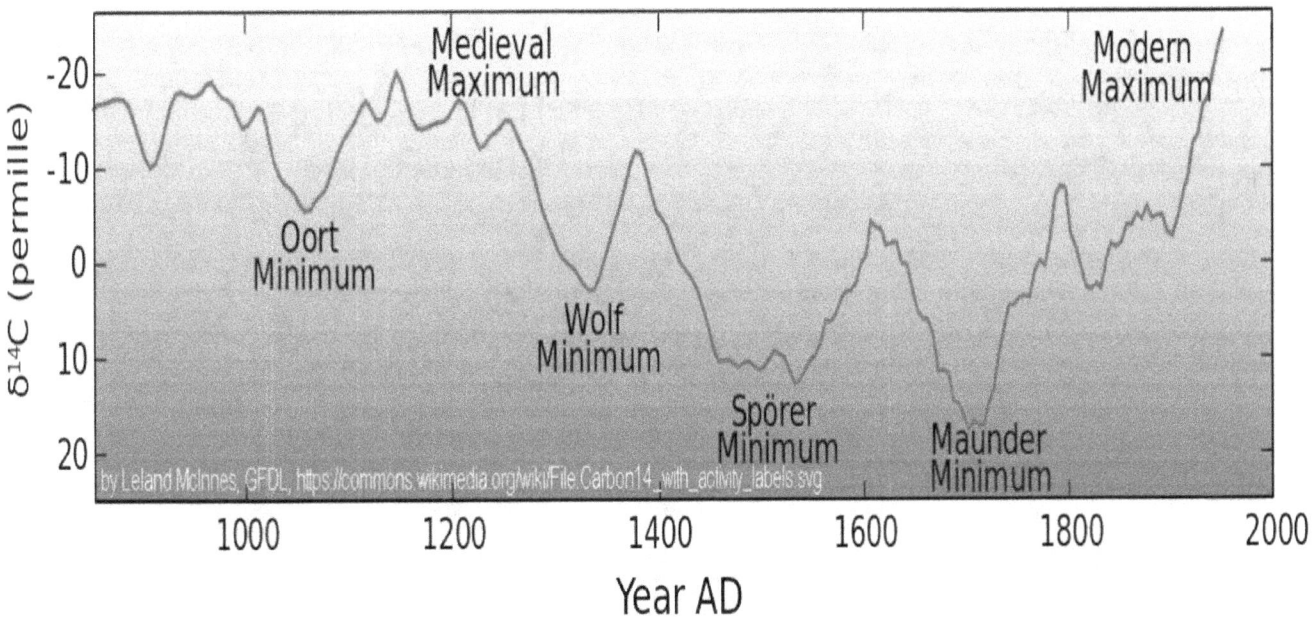

The measured and the plotted isotope ratios, like the recorded sunspot numbers, prove concretely that the Sun is a highly variable climate factor. The measured evidence indicates that when the solar activity is weak, the Earth is correspondingly cold, which is known from climate records of the respective periods.

When the Sun was ramped up in the early 1700s

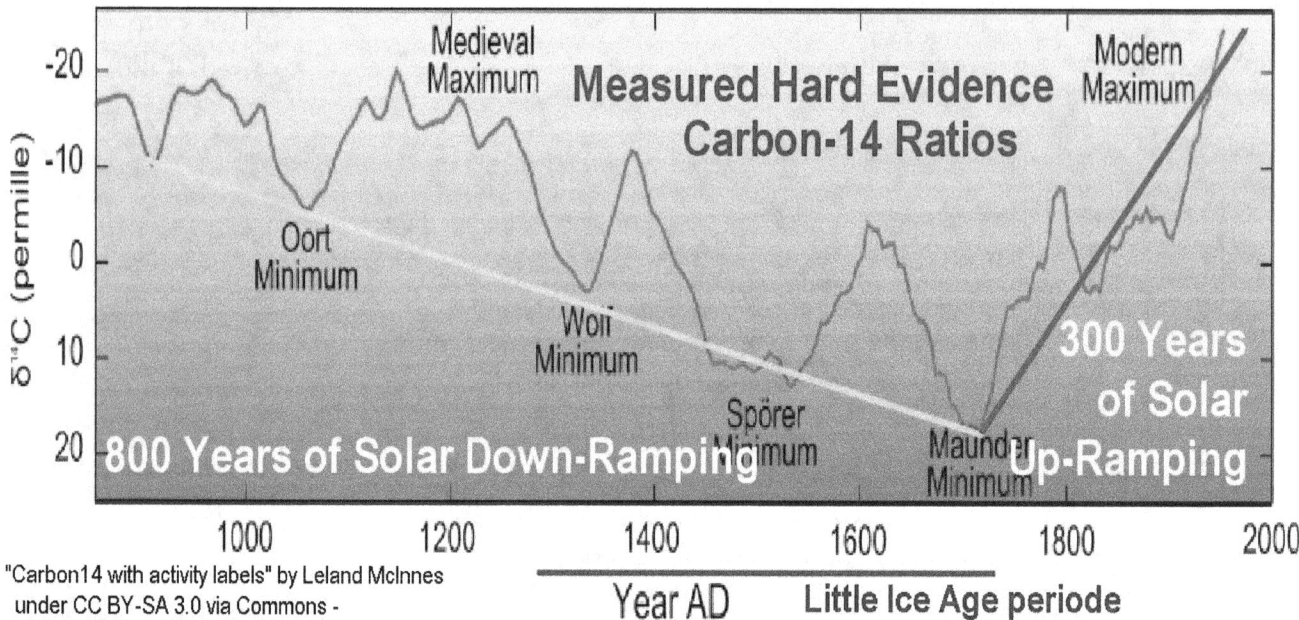

Changing solar cosmic-ray flux, measured in carbon-14 ratios, shows direct inverse relationship with known cold-climate events

"Carbon14 with activity labels" by Leland McInnes under CC BY-SA 3.0 via Commons -

 Inversely, when the Sun was ramped up in the early 1700s, the Earth became progressively warmer. The up-ramped Sun gave us a nearly 300 years of Global Warming on Earth.

The warming of the Earth, ironically, was not caused by the Sun getting hotter. It was caused by the stronger Sun emitting less solar cosmic radiation. This resulted in less cloudiness and warmer climates.

In this manner the coincidence of the historic isotope measurements with known historic climate records, prove our Sun to be tremendously changeable, and to be in fact the climate master of the Earth by changing solar activity becoming expressed in changing solar cosmic-ray flux that immensely affects our climate in a wide range of effects, such as when the global warming by the up-ramped Sun had ended the Little Ice Age.

The solar global warming had lasted all the way into the late 1990s and then ended and became reversed, when the solar activity began to diminish again. The reversal is not shown here, because it happened after the isotope ratios became invalid in 1945 in the wake of the nuclear bombing and bomb testing that added a new source for the radio-isotope production in the atmosphere.

Another radio-isotope produced by solar cosmic-ray flux

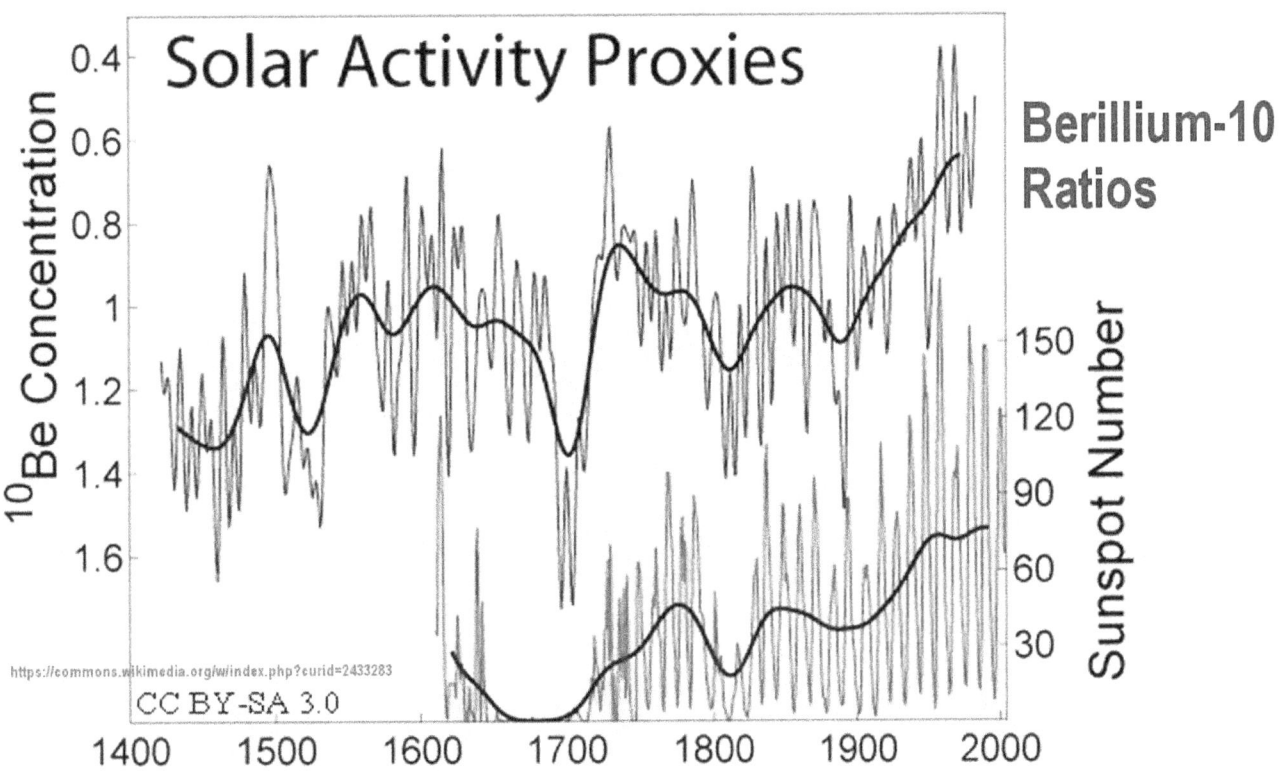

Another radio-isotope that is produced in the atmosphere by solar cosmic-ray flux, is berillium-10. The measurements of the production ratio of the berillium-10 isotope, which is preserved in historic samples, gives us one more means to measure solar activity. In this case the changing measured isotope ratio change sharply with the solar cycles, and agree remarkably close with the changing sunspot number counts.

However, the isotope ratios tell us something extremely important that the sunspot numbers don't. The isotope ratios tell us that the big climate changes on the Earth are the direct result of changing solar cosmic-ray flux that changes inversely with the intensity of solar activity. In other words, it is the solar cosmic-ray flux that is affecting the Earth and its climate in a big way.

Critical evidence that the sunspot numbers do not reveal

Critical evidence
that the sunspot numbers do not reveal

Critical evidence that the sunspot numbers do not reveal

Radiated solar energy fluctuates a mere 7/100th of a percent

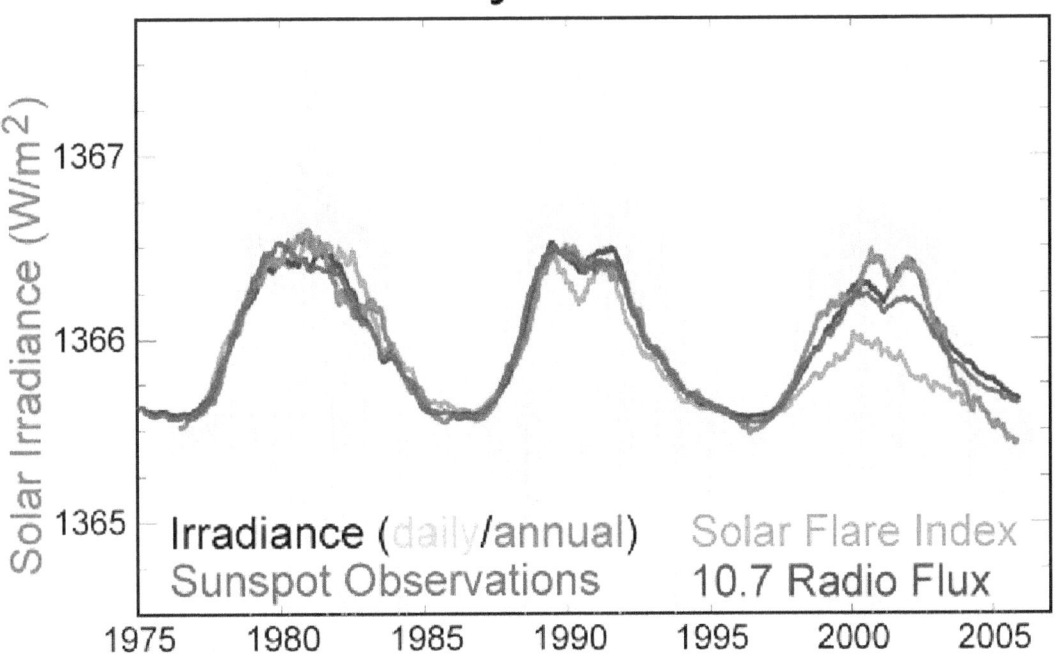

It has since been discovered by other types of measurements that the radiated solar energy does not change with changing solar activity. The energy radiation fluctuates a mere 7/100th of a percent.

The measured fact tells us that the immense cooling during the Little Ice Age periods, such as that of the 1600s in which agricultures had collapsed so extensively that 10% to 30% of the population in Europe starved to death, had been exclusively the result of very large increases of solar cosmic-ray flux affecting the Earth, that is typical for an extremely weak Sun, such as we had when no sunspots could be seen.

Cosmic-ray tremendously large in the 1600s

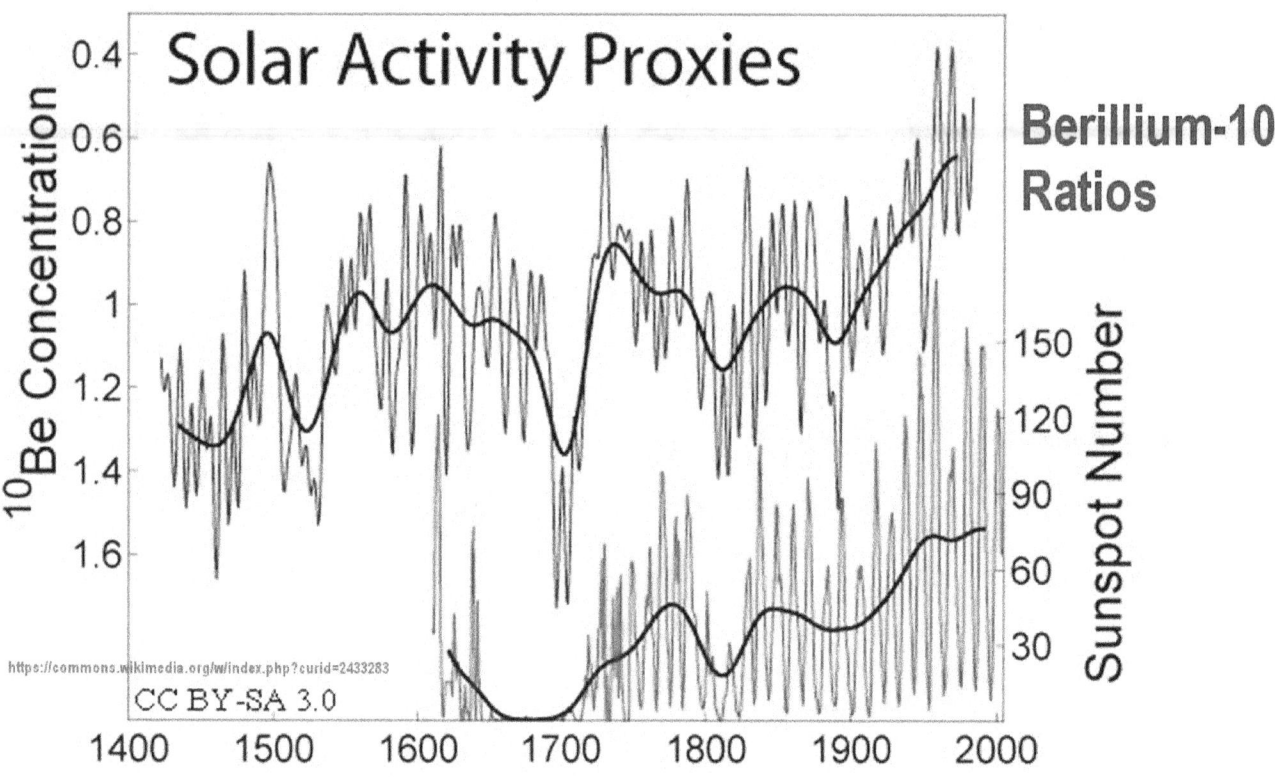

That the cosmic-ray increase was tremendously large in the 1600s is evident in the extremely large berillum-10 volume from that period.

Note: The isotope ratio is inversely plotted in this graphic so that the large isotope ratios do visibly coincide with the known historic cold climate conditions. The two deep-reaching spikes just prior to the 1700s, coincide with 30 years of no sunspots at all in the last part of the Maunder Minimum of the Little Ice Age.

Beyond Theory stands measured reality

**Beyond Theory
stands measured reality**

Beyond Theory stands measured reality

The solar forcing of our climate is not fantasy. It is measured as solar history, and it corresponds with historic climate records. The agreement is a simple case of cause and effect.

We have measured huge changes in solar activity

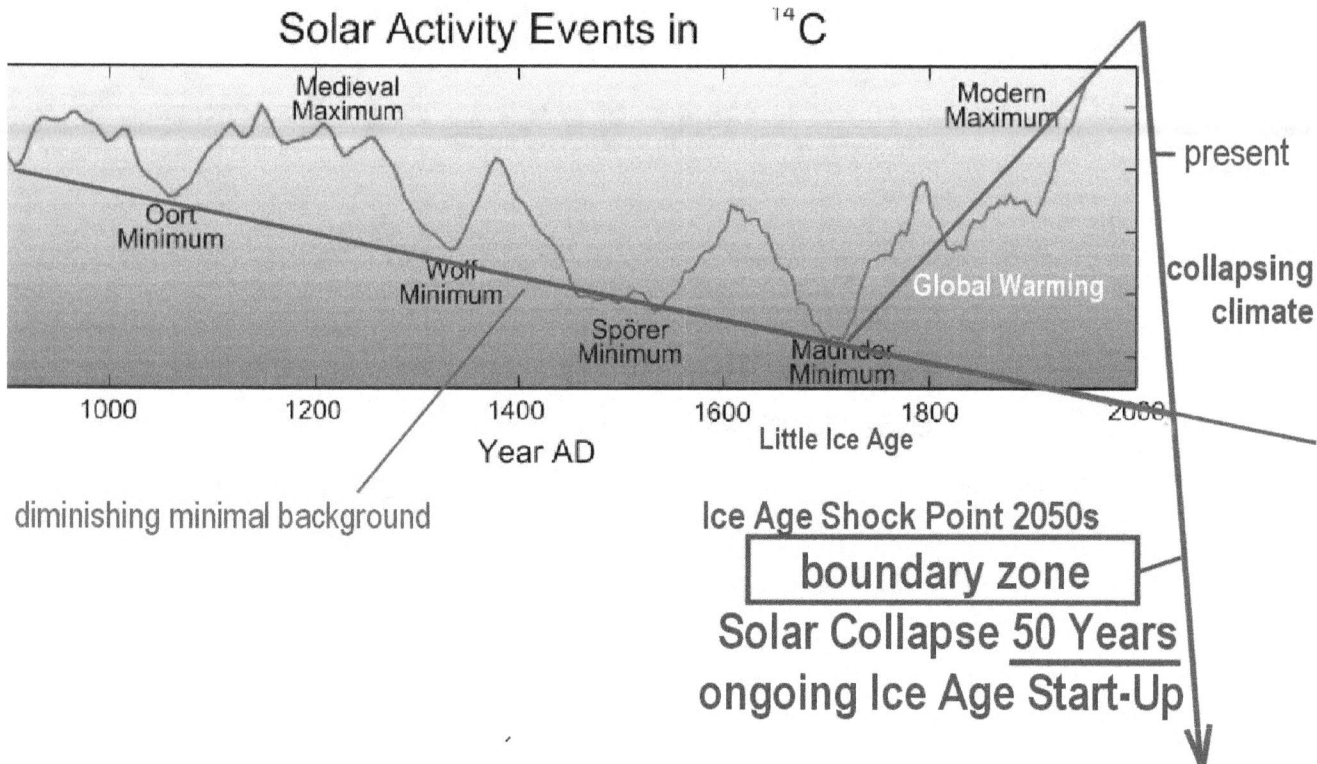

The measured solar facts are important for understanding the modern climate collapse, which needs to be understood in terms of increasing solar cosmic-ray volumes that now begin to occur under the weakening Sun, which are affecting our climate evermore in the unfolding boundary zone to the next Ice Age.

We have measured huge changes in solar activity during the last 400 years. We have measured large activity increases during the global warming period, and even larger activity decreases during the boundary zone. We have also measured corresponding changes in solar cosmic-ray flux that agree closely with the experienced changing climates. The connection of the Sun with our climate on Earth has thereby become an established fact, by real physical measurements.

Ice Age theories that ignore the Sun

**When the lights go out in the house, the house goes dark.
When solar activity diminishes, the Earth becomes colder.**

**When the lights are turned on, the house becomes brighter.
When solar activity increases, the Earth becomes warmer.**

All the climate and Ice Age theories that ignore the Sun as the causative climate factor have been rendered by the contrary measured facts to be but speculative theories without a foundation to stand on.

Part 3 - Changing Solar Activity, Measured in Real-Time

Part 3

Changing Solar Activity

Measured in Real-Time, in Proxy

Part 3 - Changing Solar Activity, Measured in Real-Time, in Proxy

New forms of measuring solar activity

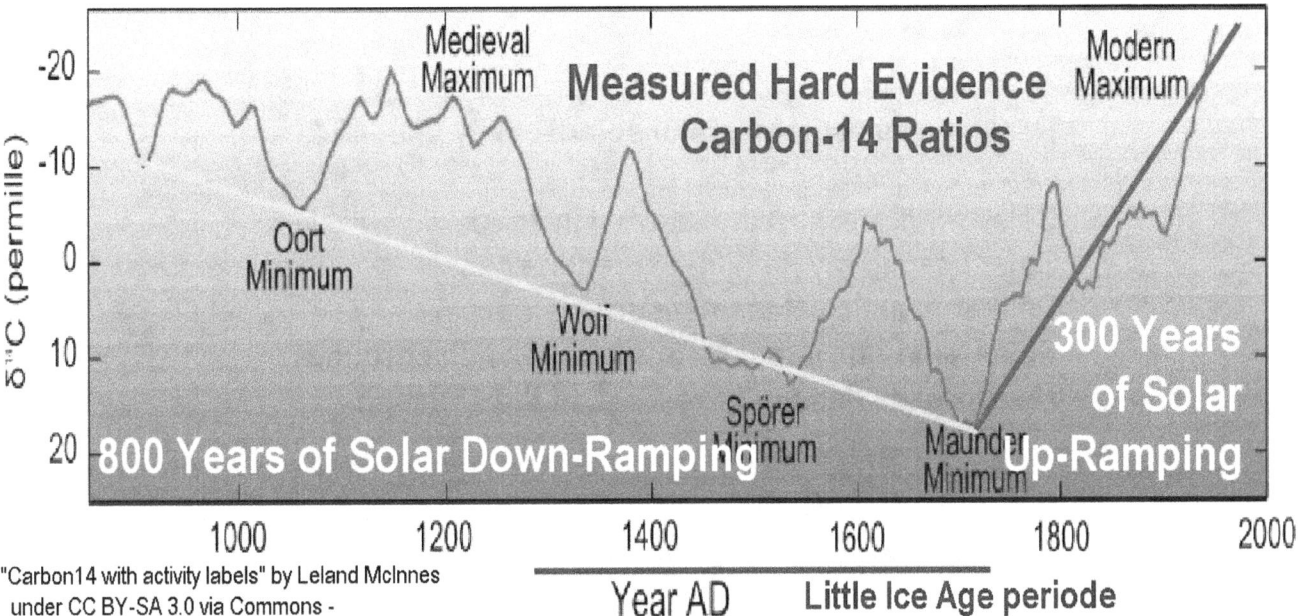

Changing solar cosmic-ray flux, measured in carbon-14 ratios,
shows direct inverse relationship with known cold-climate events

"Carbon14 with activity labels" by Leland McInnes
under CC BY-SA 3.0 via Commons -

The historically established facts didn't change in the 1940s when nuclear bomb testing, and nuclear war, created a new source for the radio isotopes that were previously produced exclusively by the Sun. The resulting loss of the measurable proxies for solar activity was overcome with a number of new forms of measuring solar activity. These new measurements, of course, were not drawn from historic samples, but were made in real time, both in space and on the ground.

Real-time measurements of solar activity after 1945

Real-time measurements of solar activity after 1945

Solar-wind pressure (measured in space)
Solar cosmic-ray flux (measured in space)
Solar magnetic fields (measured in space)
Solar radio-flux intensity (measured on the ground)
Neutron-flux density (measured in the atmosphere)

These new types of measurements do not detract from the historic measurements. They add to the proof that the Sun is changing and is massively affecting our climate, and with it our agricultures and our food production.

Principles that link climate change to solar activity

The result of these measurements takes us far out of the fantasy world of mainstream cosmology that regards the Sun as an invariable constant, and regards climate change as manmade. The real-time physical measurements reflect the principles that link climate change to solar activity.

Begun to affect agricultural food production

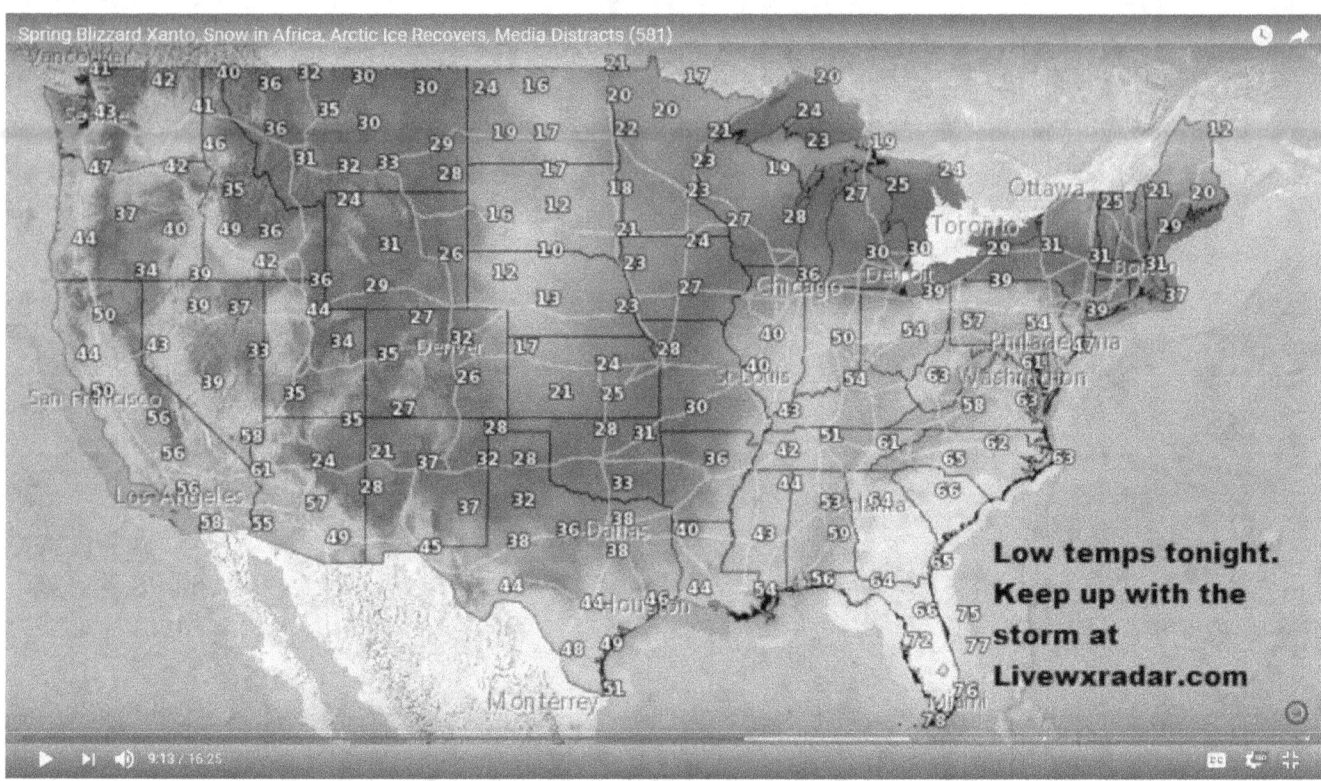

With the modern measurements now made in real time, we find the measurements reflected in real-time climate affects. Some of these now increasing real-time effects are already so extensive that they have begun to affect agricultural food production in quite a few places of the world.

In the case shown here a massive winter blizzard had blanketed the North American grain belt with snow and deep sub-zero temperatures in spring 2018, with massive snow being dumped as late as April when spring planting should have been half completed, and from which frosts may be lingering into May. Such effects, though they are not critical yet, do shorten the growing season, and may, in the years ahead, become dramatically critical.

The Sun a fast-collapsing, existentially critical, climate factor

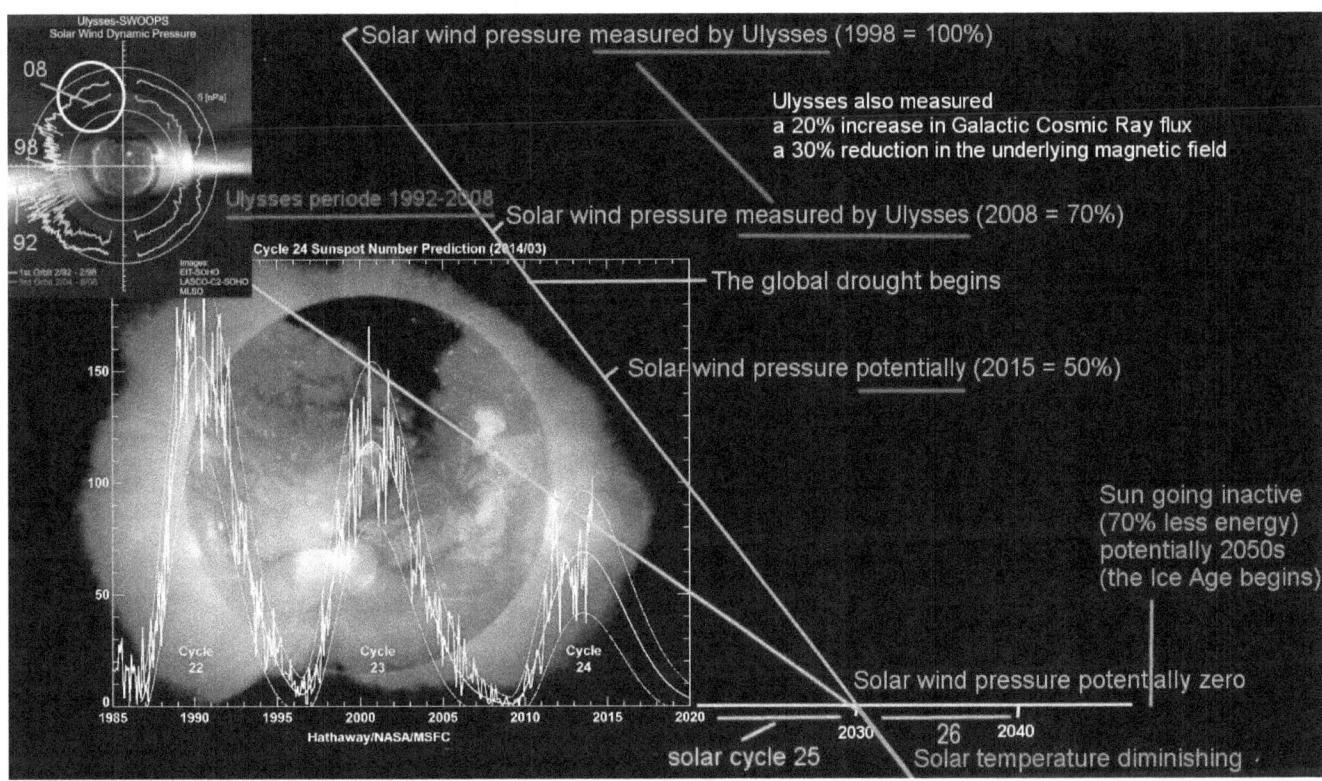

Furthermore, when these measured and experienced real-time effects are being projected forward, the Sun becomes recognized, evermore surprisingly by the unfolding dynamics as not just being merely a variable star, but as being a fast-collapsing, existentially critical, climate factor for our very living on the Earth.

Look at the modern real-time measurements

Wow! Let's back up here.

Let's take a closer look at the modern real-time measurements and what they are telling us.

Wow! Let's back up here. Let's take a closer look at the modern real-time measurements and what they are telling us.

Ulysses spacecraft had measured the Sun directly

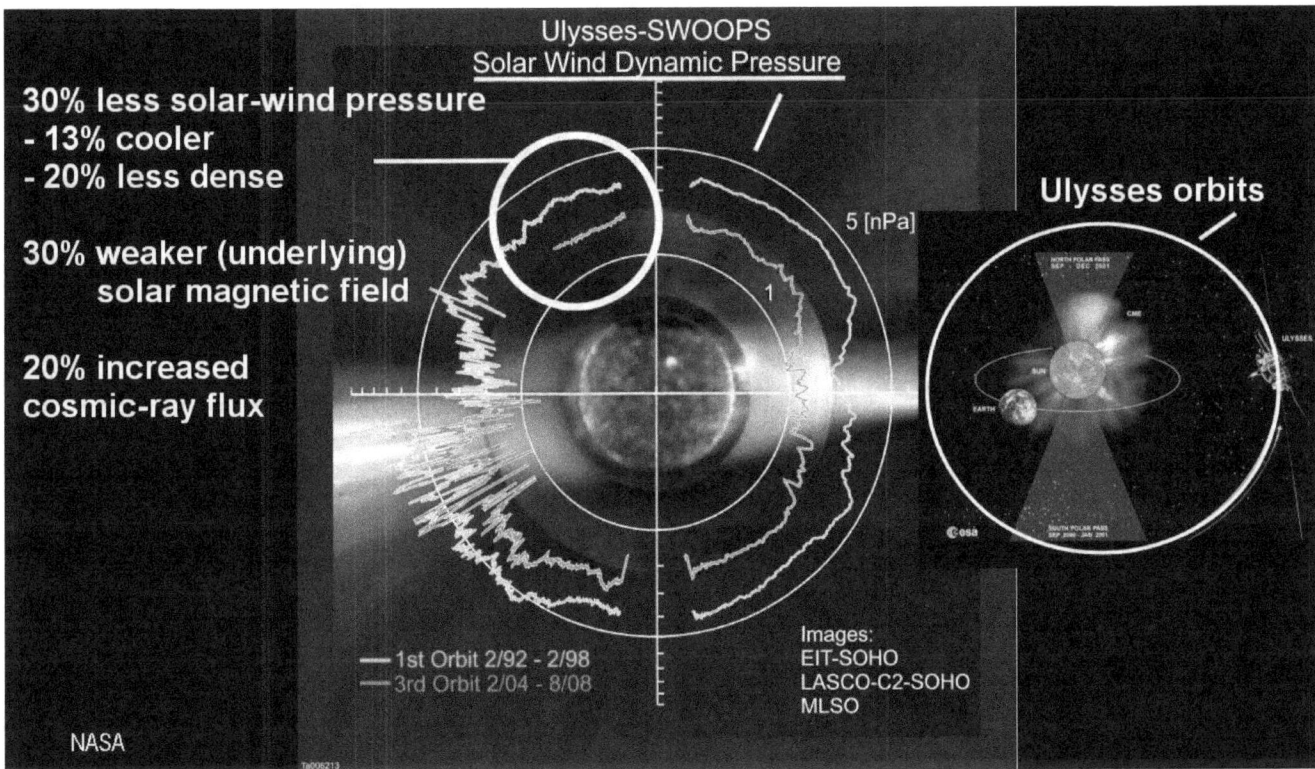

Two of the new types of physical measurements that were conducted in space, were conducted by the NASA and ESO Ulysses spacecraft that had measured the Sun directly from a wide polar orbit. It measured the solar activity as it happened, and 'saw' it diminishing, and not just a little. Its measurements indicate that the solar activity, in terms of solar wind pressure, was diminishing at the amazingly fast rate of 30% per decade from the late 1990s onward, till 2008 at the end of the mission.

Enormous dropping off in solar activity

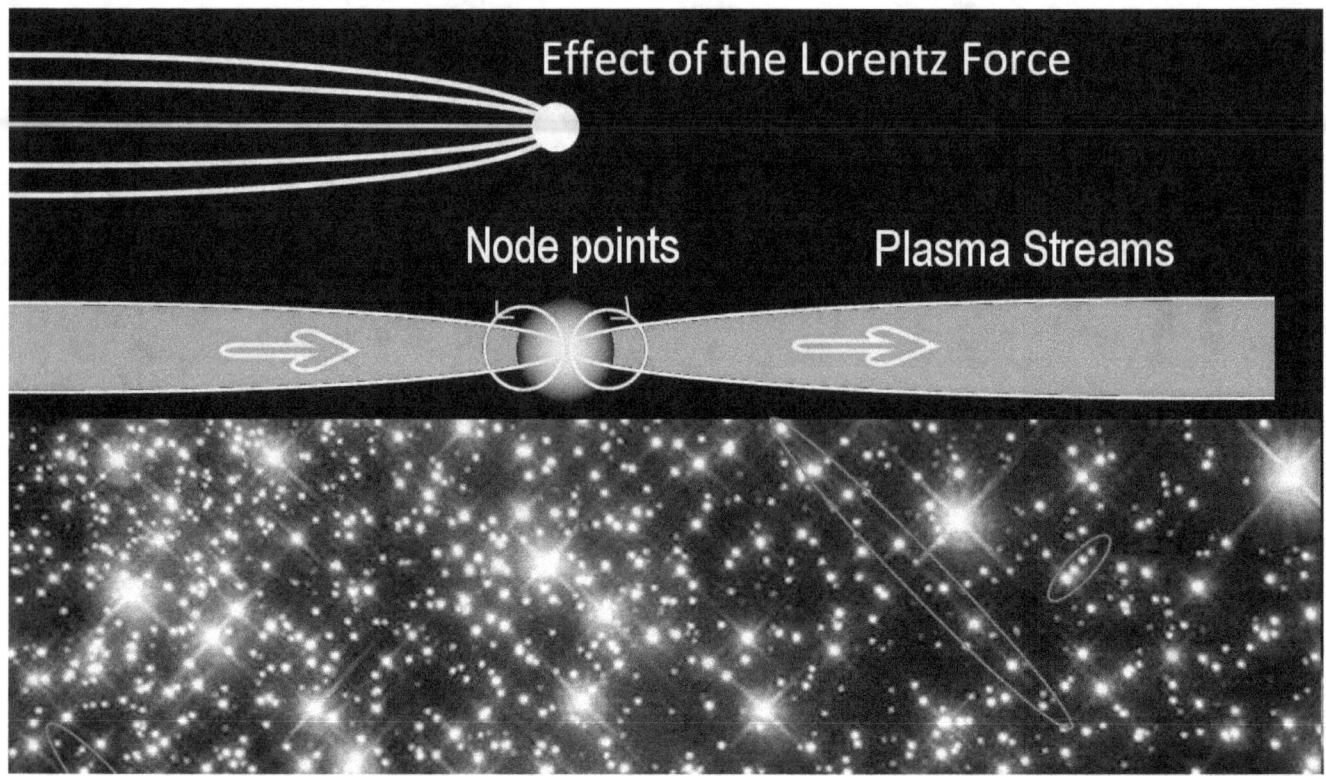

This enormous dropping off in solar activity is possible by the fact that the Sun is located within a complex dynamic system of interstellar plasma streams that power its operation. These plasma streams are focused onto the Sun by electromagnetic primer fields that form naturally by simple physical principles.

As the concentrated plasma around the Sun, fluctuates

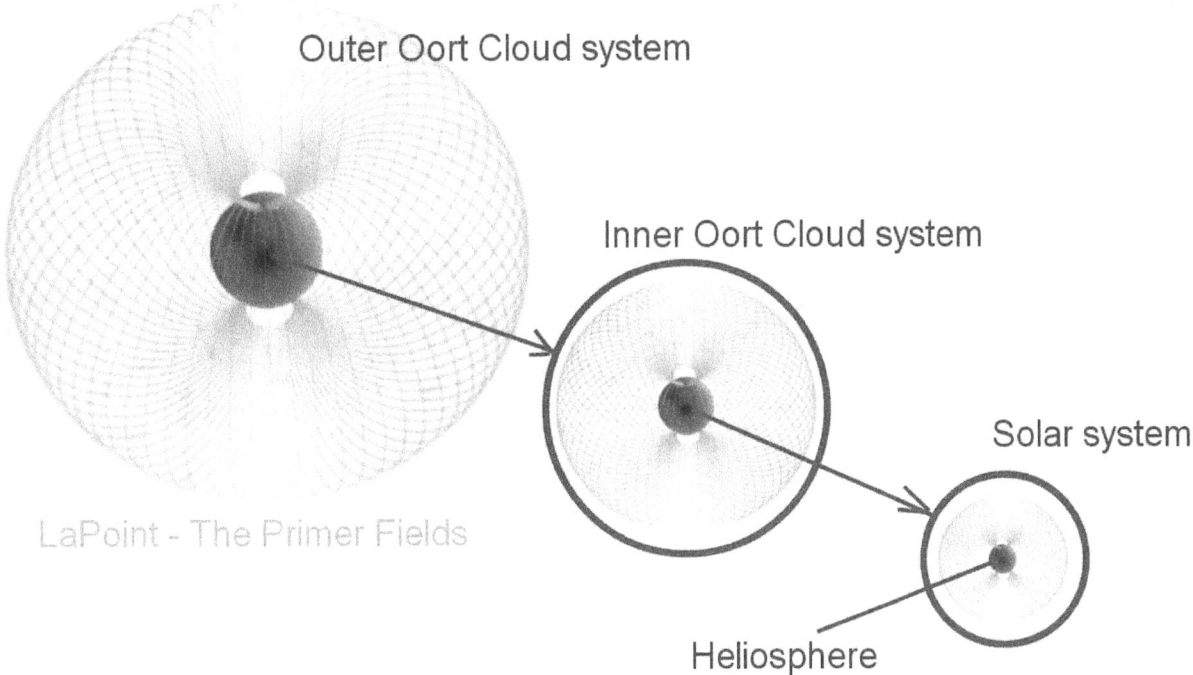

As the concentrated plasma that is built up around the Sun, fluctuates in density as the result of numerous resonance effects within the system, the solar activity fluctuates accordingly. That's the principle.

The Little Ice Age was created and terminated by the same principle

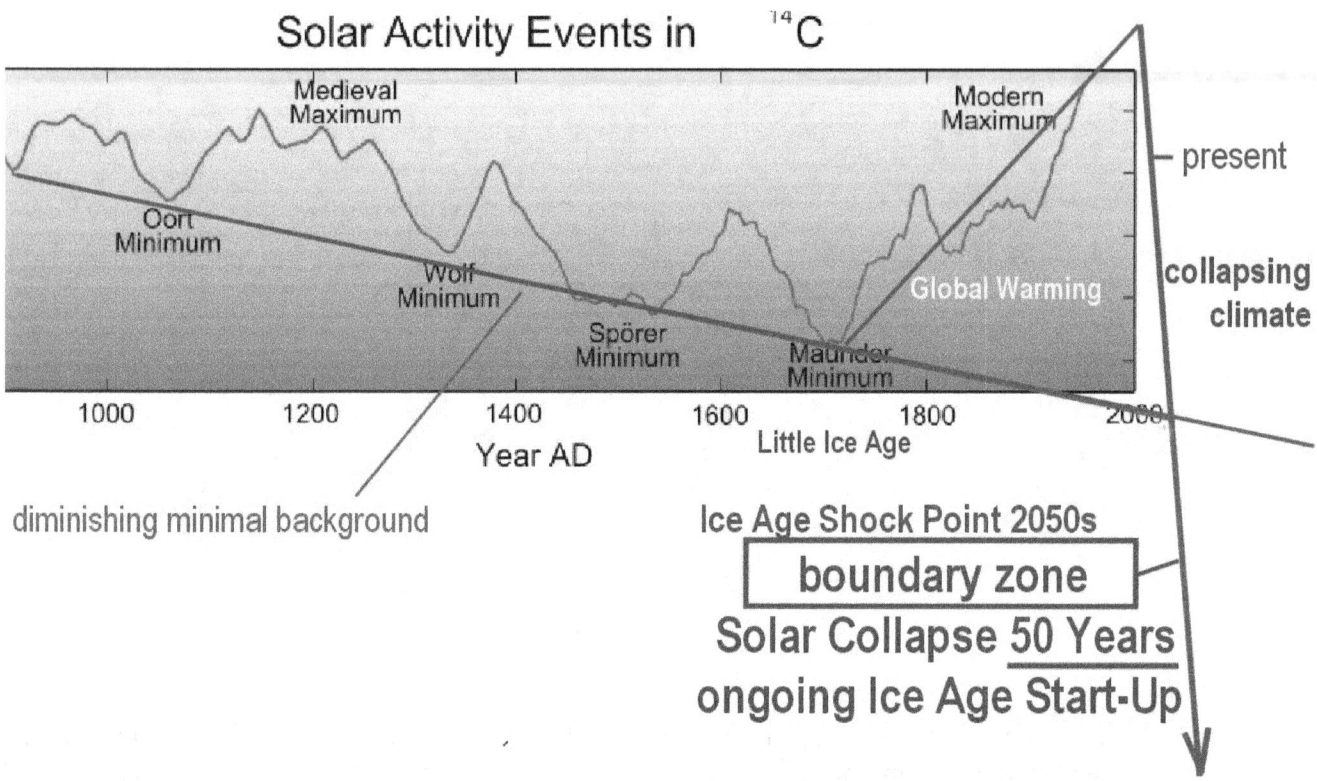

The Little Ice Age was created on this basis, and it was terminated by global warming on this same basis, which in turn became reversed again by the now on-going collapse of solar activity - all by the same principle.

The solar surface temperature will diminish

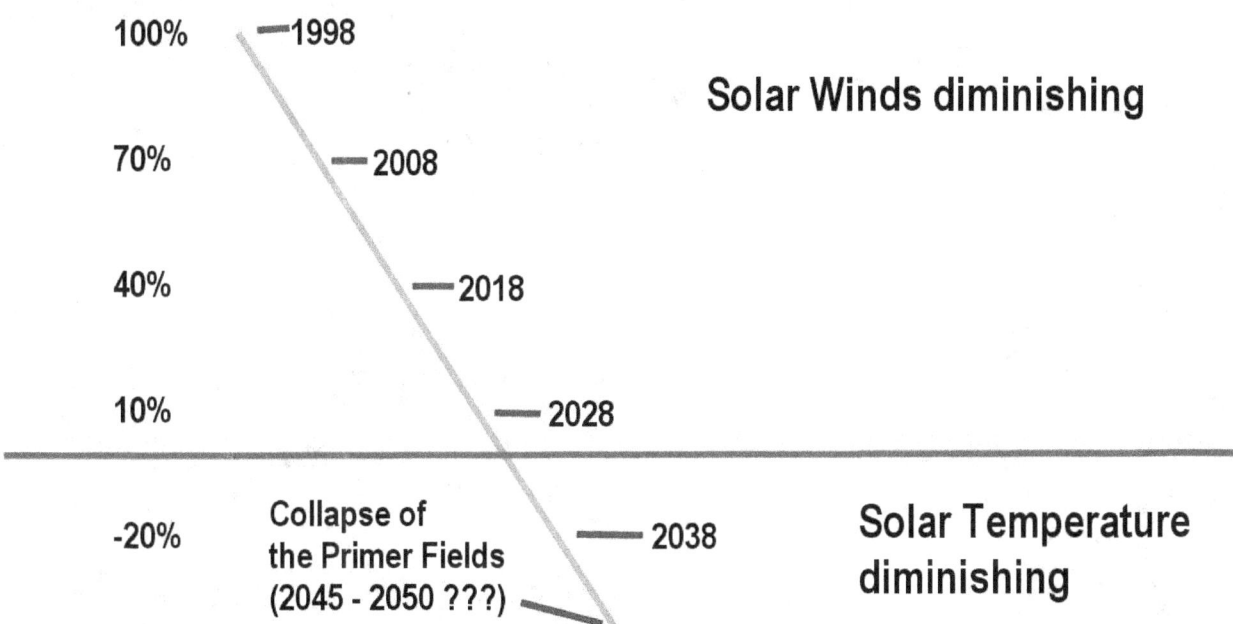

If one projects this rate of collapse of the solar-wind pressure forward in time, the interstellar powering of the Sun will have diminished to the point that the solar wind that is flowing from the Sun stops in the early 2030s. When the powering of the Sun diminishes further, the solar surface temperature will necessarily begin to diminish as the next 'shoe' begins to drop. That's big!What we see happening here are enormously large effects. Now let's considering that we are presently only half-way down that slope, even while the climate effects are already becoming 'monsters' in some cases.

Ulysses had also measured the solar cosmic-ray flux

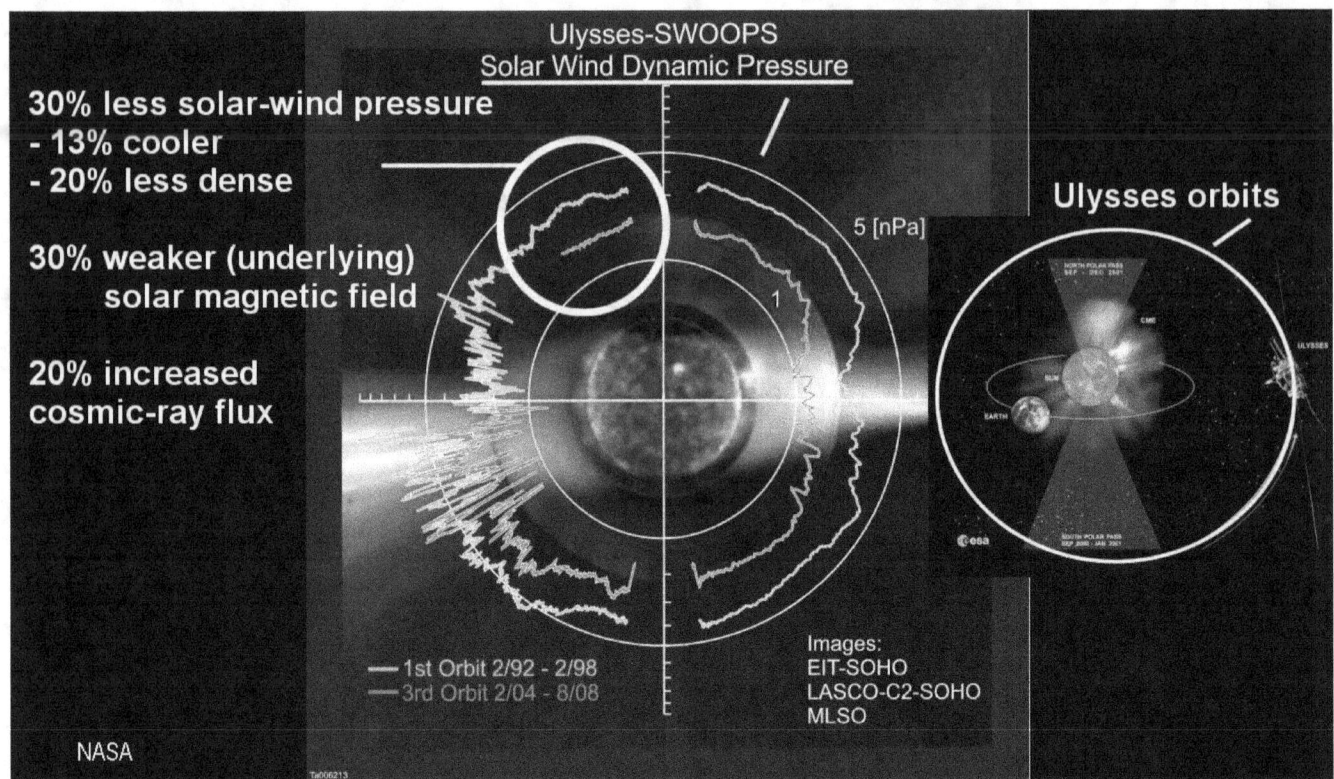

The Ulysses spacecraft had also measured the solar cosmic-ray flux during its mission, and it 'saw' it increasing by 20% over the same timeframe, and it further measured the underlying solar magnetic field, which it saw diminishing by 30%. These too, are enormous changes in solar activity, which too, are bound to have corresponding climate effects.

Ulysses 'flew' three orbits around the Sun

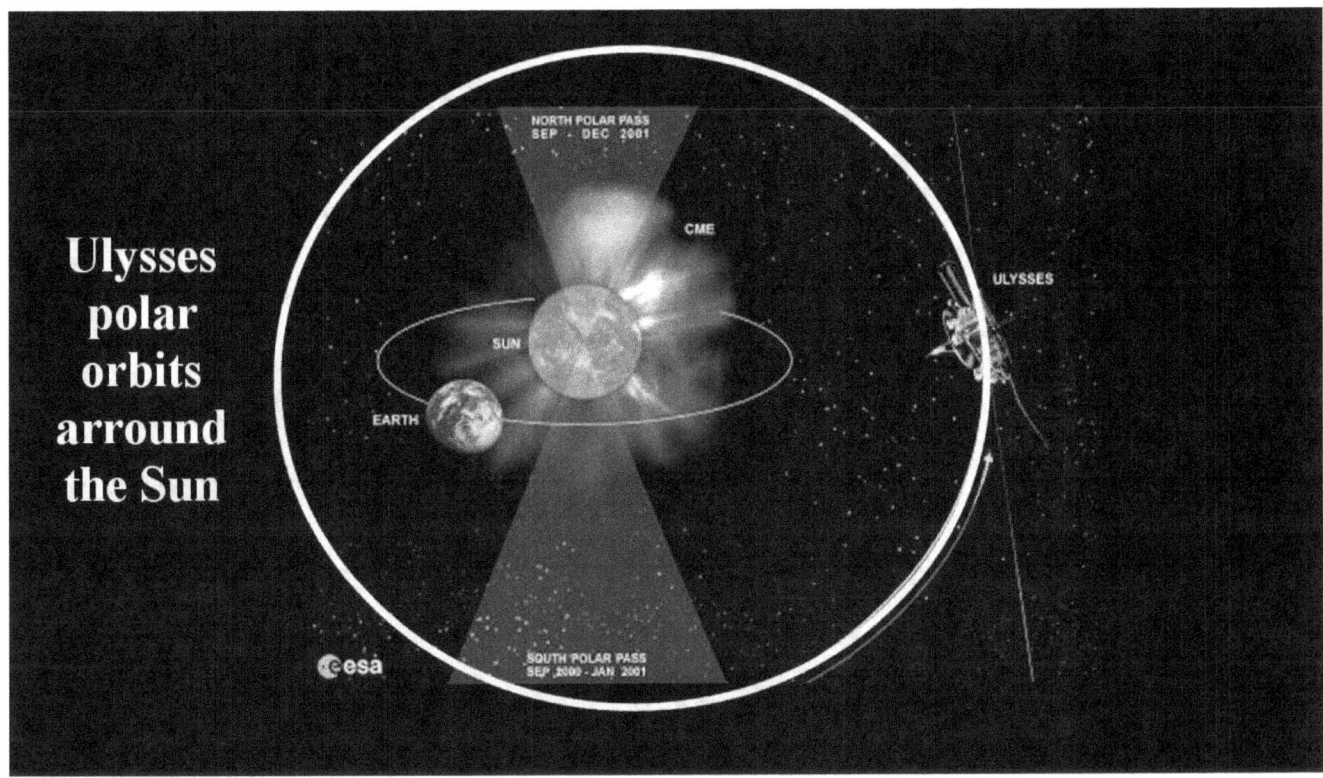

Ulysses 'flew' three orbits around the Sun, from 1995 to 2008, which covered in time the entire solar cycle 23, from start to finish. The solar wind pressure had collapsed 30% over this timeframe.

Rate of collapse continued after the Ulysses mission

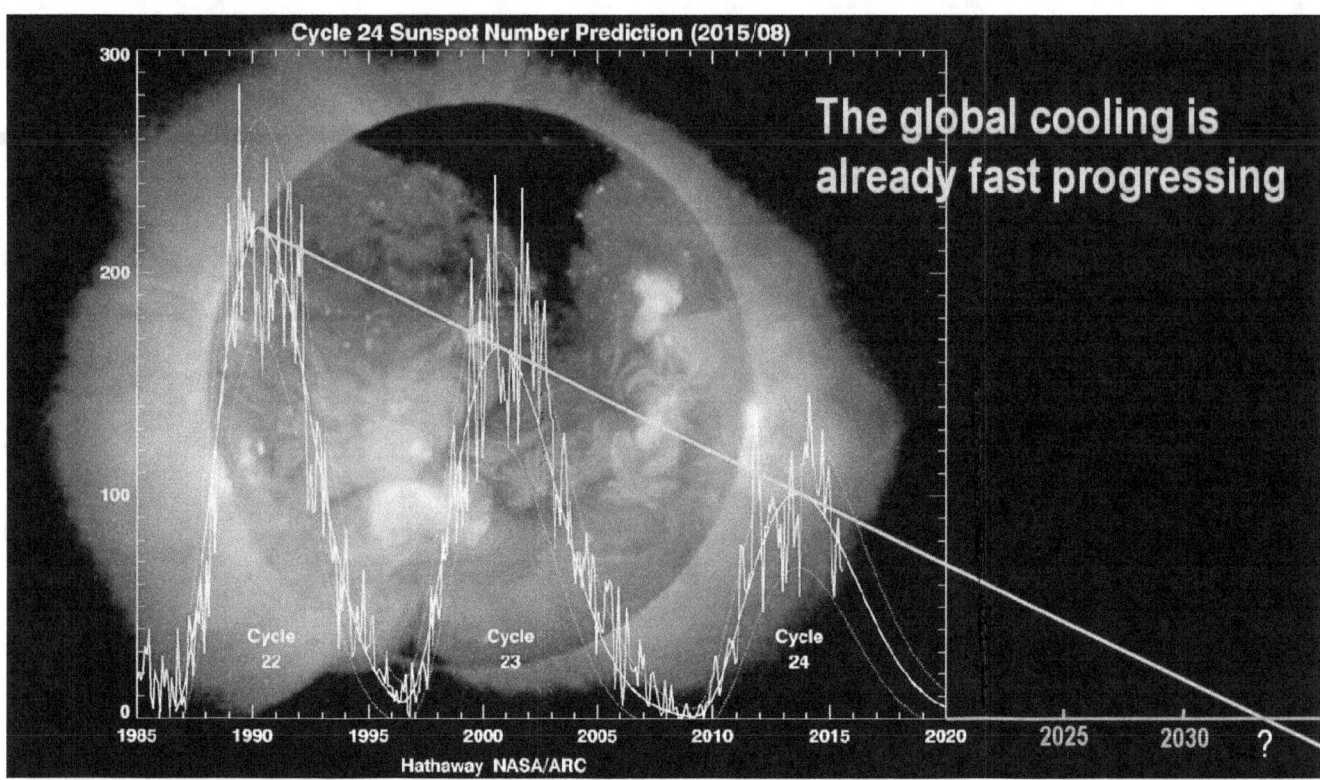

The 30% rate of collapse is also reflected in the diminishing solar cycles themselves, which evidently began before and continued after the timeframe of the Ulysses mission.

Solar radio observatories in the 10.7 cm band

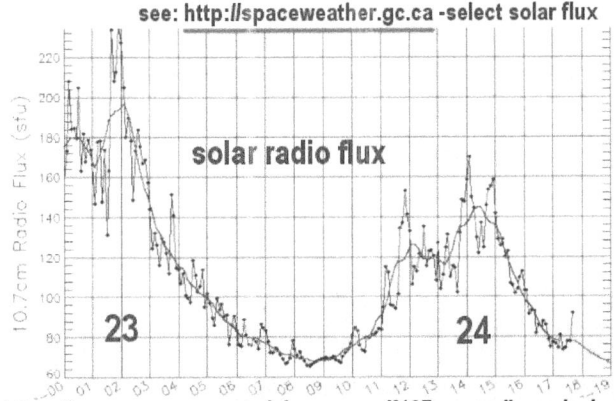

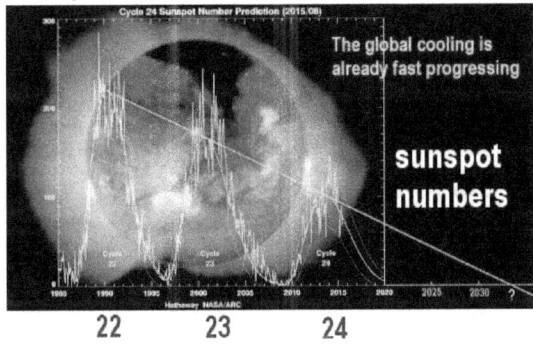

By Jjnishiyama at English Wikipedia, CC BY 2.5,
https://commons.wikimedia.org/w/index.php?curid=41186116

Other types of measurement likewise indicate that the solar activity continued to diminish past the termination of the Ulysses mission, and continues so all the way to the present. This too is physical reality measured in real time.

One of the continuing measurements is conducted by solar radio observatories in the 10.7 cm band. The 10.7 cm Solar Flux is currently one of the best indices of solar activity that we have. It covers more than 50 years of direct measurements. Only the sunspot number counts cover a longer period. In comparing the two, a linear relationship between the 10.7cm Solar Flux and the sunspot numbers becomes evident.

The dramatic collapse of the solar activity that Ulysses had observed, between the start and end of solar cycle 23, is seen to continue on in the diminishing solar radio-flux measurements in the 10.7 centimetre band. The measurements went way above 200 radio flux units at the peak of previous solar cycles, but for the peak of the last solar cycle, cycle 24, the radio solar observatories measured only half of the historic peak flux. A researcher, in referring to these measurements, commented, "the Sun is dying."

The researcher wrote

The researcher wrote,

"because the Sun is operating at half power already you can expect the Sun dying when the full solar minimum hits in 2024. The Sun/Earth will not recover from this.

From this year (2018) forward, every year will become colder. Leave Europe within 3 years, before all borders close. "(Marc C.)

The researcher wrote,

"because the Sun is operating at half power already you can expect the Sun dying when the full solar minimum hits in 2024. The Sun/Earth will not recover from this. From this year (2018) forward, every year will become colder. Leave Europe within 3 years, before all borders close.?"(Marc C.)

Comment justified by the measurements?

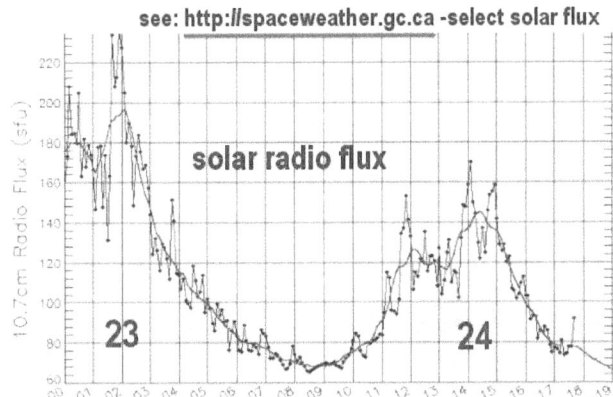

see: http://spaceweather.gc.ca -select solar flux

solar radio flux

23   24

https://www.swpc.noaa.gov/phenomena/f107-cm-radio-emissions

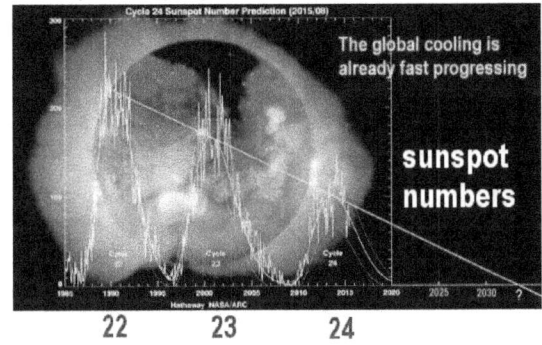

The global cooling is already fast progressing

sunspot numbers

22   23   24

Dominion Radio Astrophysical Observatory's 10.7cm solar flux monitoring telescope

Is this type of comment justified by the measurements? Just look at the latest measured ratio. At the peak of the current solar cycle, in 2014 - a weak cycle to begin with - the solar flux was measured as 170 units. Now in 2018, a mere 4 years later, the solar flux was measured at 68.9 units. That's a 60% drop. That's huge for such a short time, and more so if one considers that we are still potentially as much as 6 years away from the lowest point of the current solar cycle.

(see: http://spaceweather.gc.ca -select solar flux)

Does it mean, the sky is falling?

What does it all mean?
Does it mean, the sky is falling?

No, the world is not in a crisis yet.

What does it all mean? Does it mean, the sky is falling? No, the world is not in a crisis yet.

The boundary zone divided into three consecutive phases

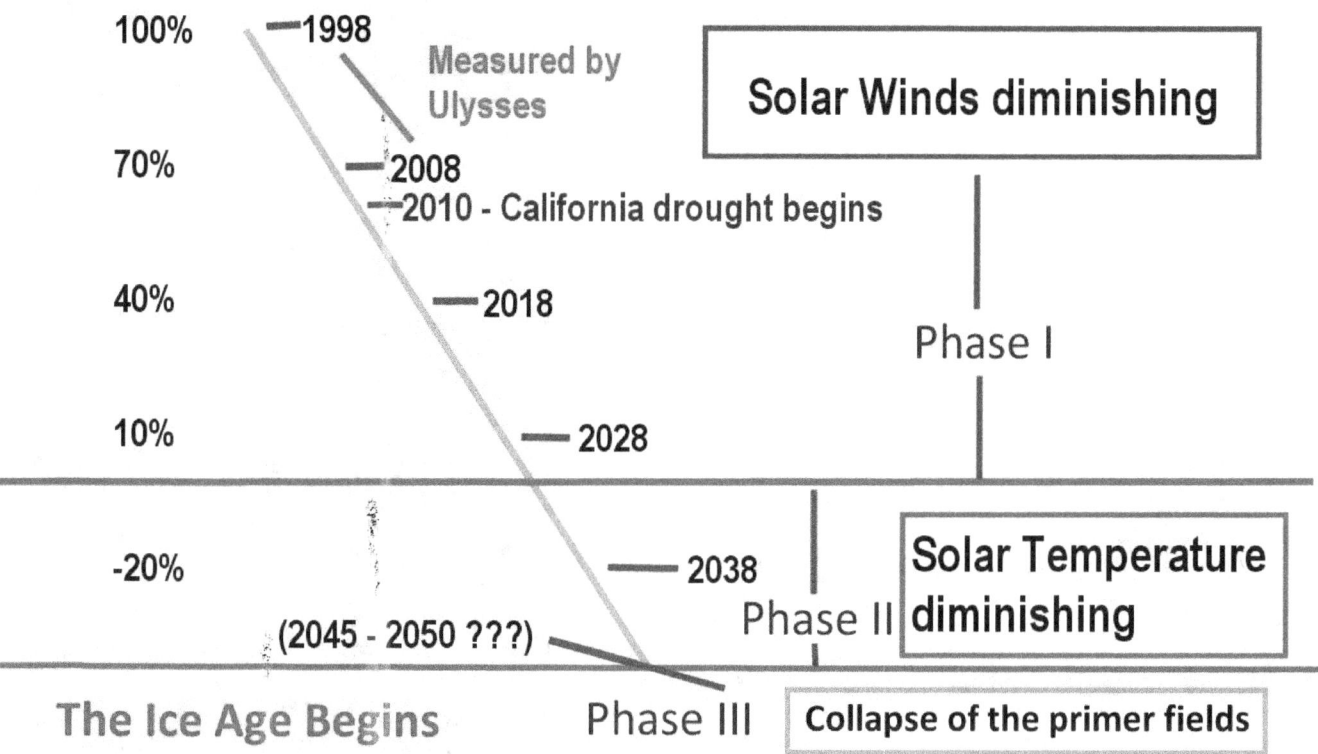

The boundary zone to the next Ice age covers a lengthy period that can be divided into three consecutive phases.

The 1st phase is the phase in which the weakening of the Sun - of the entire system that powers the Sun - becomes expressed in diminishing solar-wind pressure, while the Sun's surface temperature remains constant.

How is this possible?

Near constant solar surface temperature is possible

Maximum temperature
of liquid water
at ambient pressure
is 100 degrees Celsius:
The Boiling Point

The near constant solar surface temperature of the diminishing Sun is possible, because the solar wind is comparable to a boiling kettle venting off steam.

It is not possible to heat water in a kettle hotter than 100 degrees at sea-level altitude. When more heat is applied to the kettle, the excess heat converts water into steam, which is vented into the air.

Excess plasma pressure is simply vented back into space

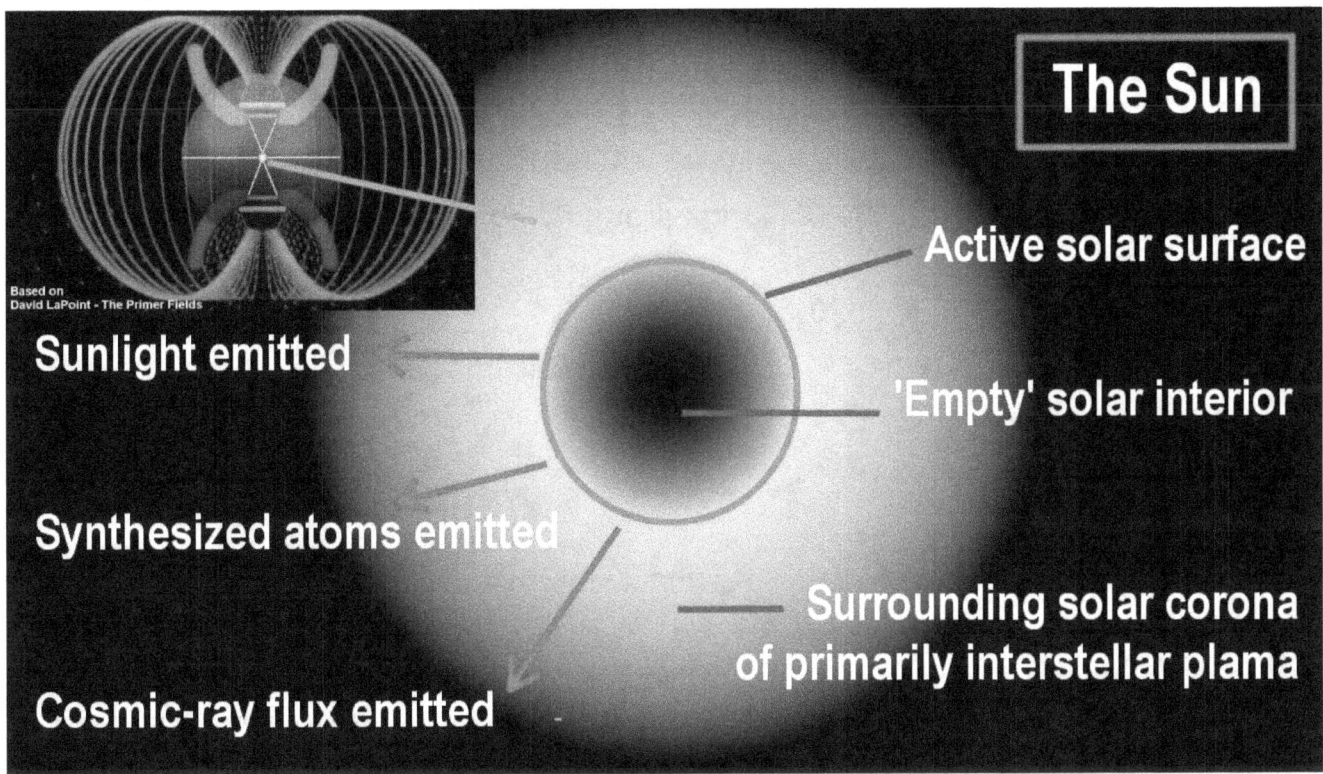

In a similar manner on the Sun, when the plasma pressure onto the Sun - which is its energy input - exceeds what the fusions cells on its surface require for the plasma-fusion process, the excess plasma pressure is simply vented back into space and becomes the solar wind. This means that for as long as the solar wind keeps flowing, there is enough plasma density in the system to keep the Sun 'fully' powered. This also means inversely, that when the solar wind is diminishing, the entire dynamic system that powers the Sun, which focuses plasma unto the Sun, is getting weaker, whereby less solar wind is vented.

Every change in solar activity affects the climate on Earth

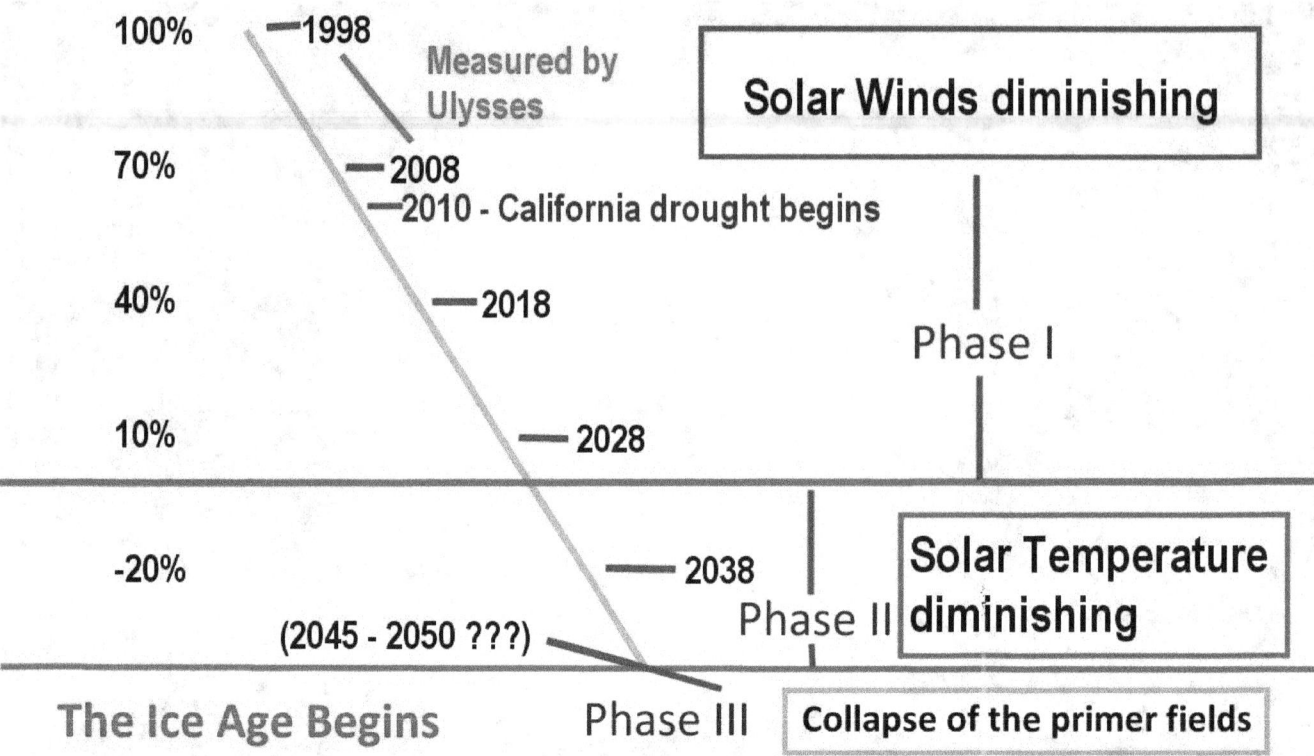

This type of weakening happens throughout Phase-1 in the boundary zone. The radio-flux that is measured with radio telescopes, reflects in general the intensity of the active processes by which the solar wind is being generated. The diminishing solar wind, tells us that the entire active system is getting progressively weaker, and that the escaping solar cosmic-ray flux, that affects the climate on Earth, is getting larger, with the climate effects getting larger too.

Every change in solar activity affects the climate on Earth. This influence is almost exclusively the result of the effects caused by solar cosmic-ray flux in the Earth's atmosphere. The influence is dramatic, and it increases with diminishing solar activity.

Cosmic-rays are fast moving, high-energy plasma particles

© Milloslav Druckmuller/Barcroft

http://www.zam.fme.vutbr.cz/~druck/Eclipse/   - an example of the amazing solar eclipse photography of Milloslav Druckmueller

The solar wind itself does not have a significant effect on the climate on Earth. Its is slow moving plasma, in the order of 800 Km/sec., while cosmic-rays are fast moving, high-energy plasma particles with enough punch to have that major effect on cloud nucleation.

Cosmic-rays are single events

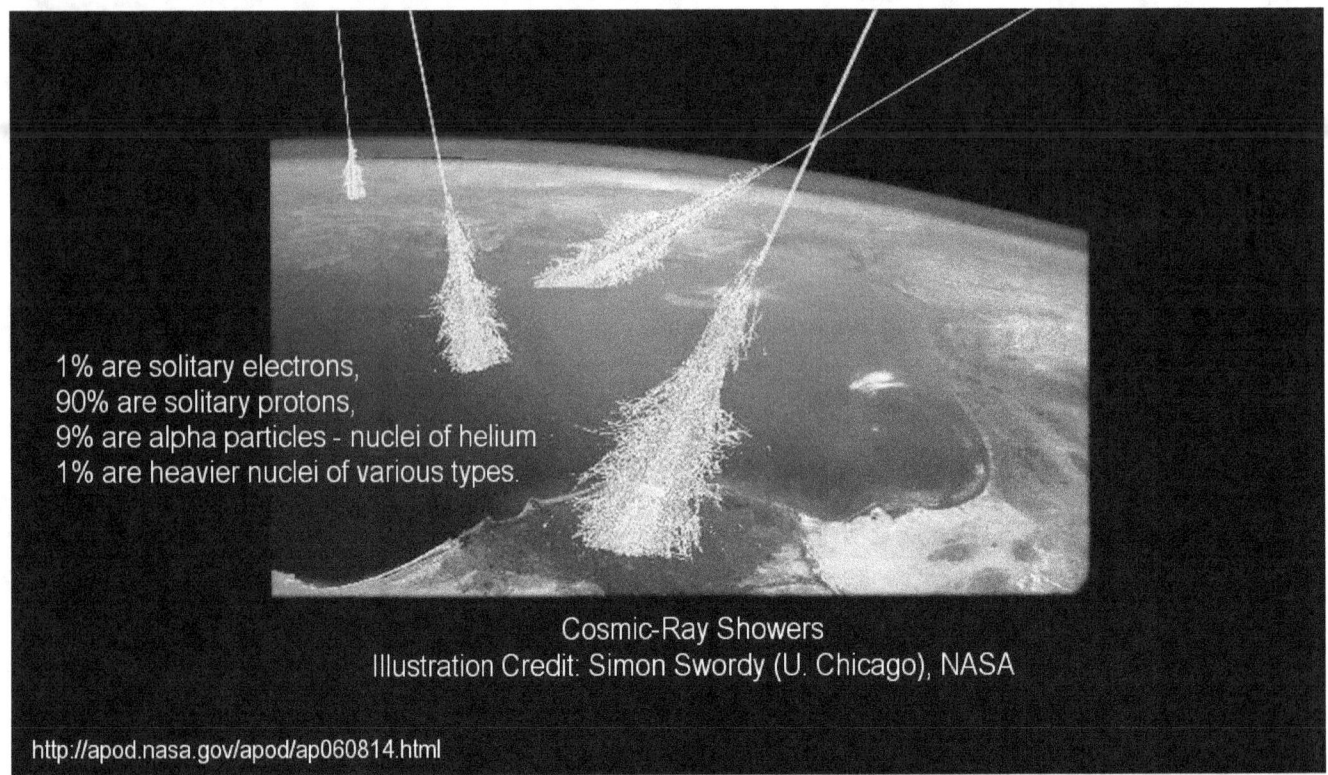

1% are solitary electrons,
90% are solitary protons,
9% are alpha particles - nuclei of helium
1% are heavier nuclei of various types.

Cosmic-Ray Showers
Illustration Credit: Simon Swordy (U. Chicago), NASA

http://apod.nasa.gov/apod/ap060814.html

Cosmic-rays are single events of highly-energized plasma particles, primarily protons, that escape the solar fusion cells and then penetrate the plasma sphere around the Sun where most of the escaping plasma particles become trapped. The few that are not trapped escape into space as solar cosmic-ray flux. When the escaping cosmic-ray flux interacts with the Earth's atmosphere, it affects the cloud-forming process enormously, by way of cosmic-ray ionization. Ionized air is 100-times more attractive for the cloud nucleation process than non-ionized air. As the result of the ionization, cloud aerosols form more readily, and condense more readily into water droplets, by which the clouds rain out.

As a consequence of increased cloudiness

As a consequence of increased cloudiness, the Earth is getting colder. The white top of the clouds reflect a portion of the incoming sunlight back into space. More clouds reflect more sunlight away. The reflected energy is lost to us. The Earth invariably becomes colder in an accumulative process. This type of increased cooling by the weakening Sun is evermore evident in recent times all over the world in the forms of increased snowfall, frosts, late blizzards, and so on.

Increased rain-out reduces the water transport distance

wikipedia - by Muhammad Mahdi Karim

The increased rain-out of the clouds also reduces the water transport distance by the clouds. Droughts are formed by this process, and in some cases the enhanced rain-out also causes floods.

When a major cosmic-ray shower erupts

The lower portion of the troposphere may contain as much as 4% of its volume in the form of water vapor. If all of that was turned to water, the land would be covered 100 feet deep in water. Normally only a minuscule portion condenses into clouds. But when a major cosmic-ray shower erupts, typically through coronal holes on the Sun, cloud nucleation can be dramatically increased and generate flash floods right in the midst of a drought.

After years of drought, flash floods erupted

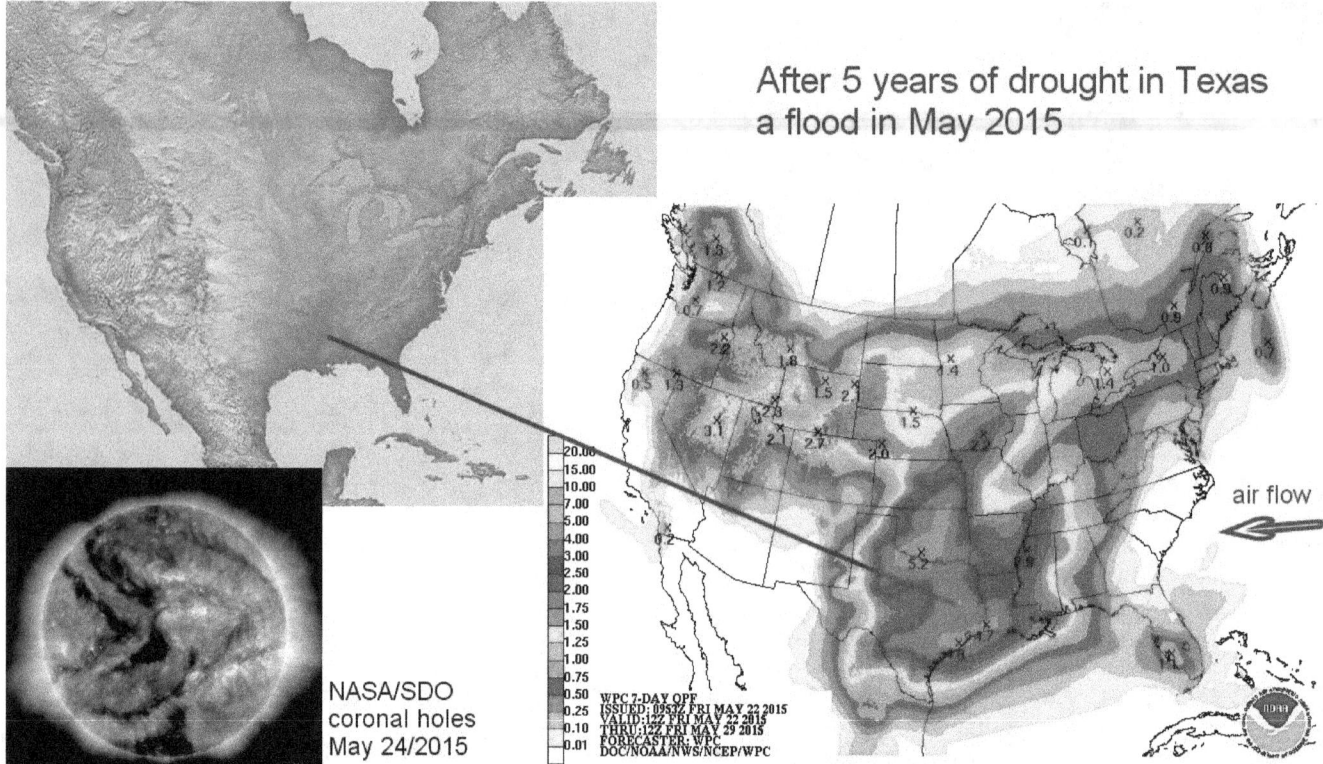

After 5 years of drought in Texas a flood in May 2015

NASA/SDO coronal holes May 24/2015

air flow

WPC 7-DAY QPF
ISSUED: 0953Z FRI MAY 22 2015
VALID:12Z FRI MAY 22 2015
THRU:12Z FRI MAY 29 2015
FORECASTER: WPC
DOC/NOAA/NWS/NCEP/WPC

Such an event occurred in Texas in 2015. After years of drought, flash floods erupted from Texas all the way north to Canada. These types of effects are typically produced by large solar cosmic-ray events.

This is how changing solar activity becomes expressed in large climate anomalies on Earth, because the anomalies are caused by the effects of changing cosmic-ray flux. In fact, changing cosmic-ray flux is presently the only driver for climate changes on Earth, with a few minor exceptions. Enormous climate effects can be caused by changing solar cosmic-ray flux while the Sun's surface temperature remains solidly constant.

Part 4 - Climate Change by Solar Cosmic-Ray Flux

Part 4

Climate Change

Forced by Changing Solar Cosmic-Ray Flux
almost exclusively

Part 4 - Climate Change: Forced by Changing Solar Cosmic-Ray Flux - almost exclusively

Solar cosmic-ray flux the only active climate factor

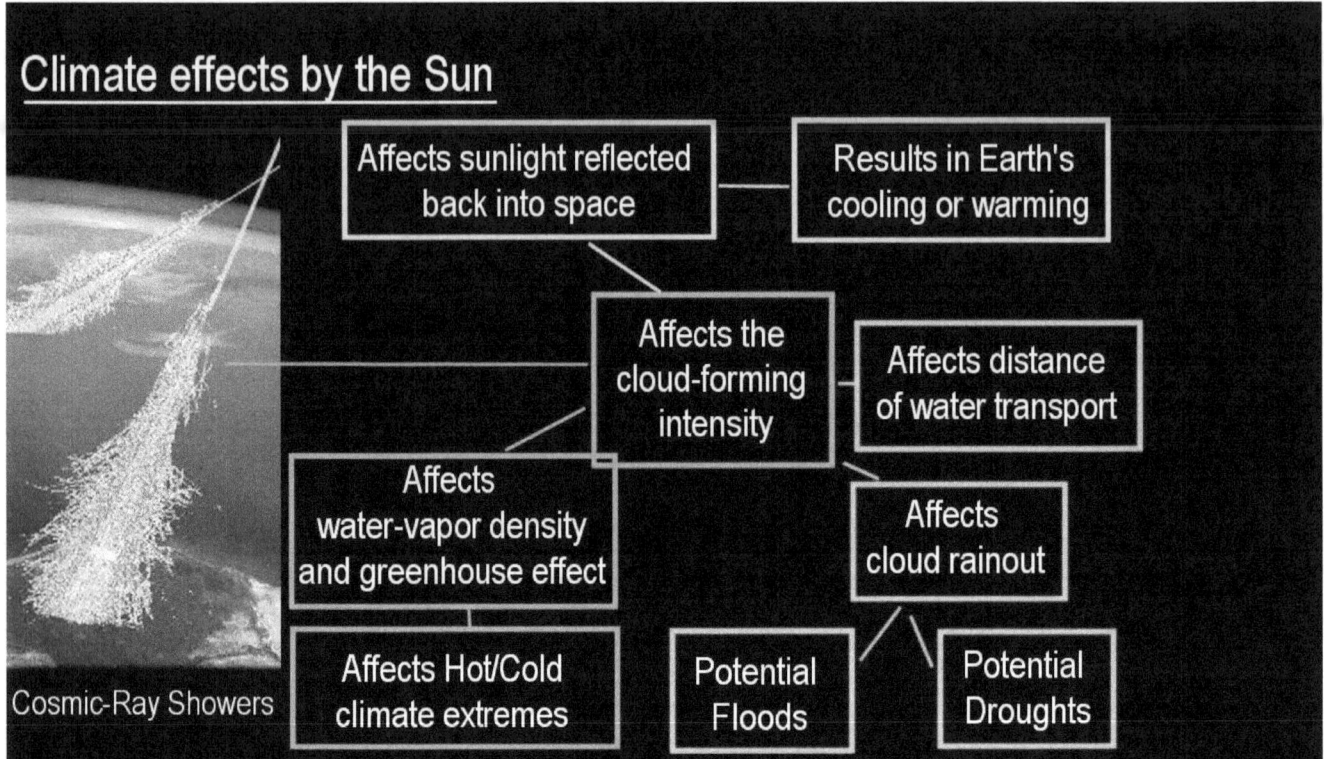

This means that the solar cosmic-ray flux is presently the only active climate factor. And as I said before, this factor affects us almost exclusively by its effect on the cloud forming process. And as I said also, changing cloudiness affects the entire spectrum of climate effects from hot to cold, from drought to flooding, and it affects even the moderating effect of the Earth's greenhouse mantle that is caused over 90% by water vapor.

So, how do we know all this?

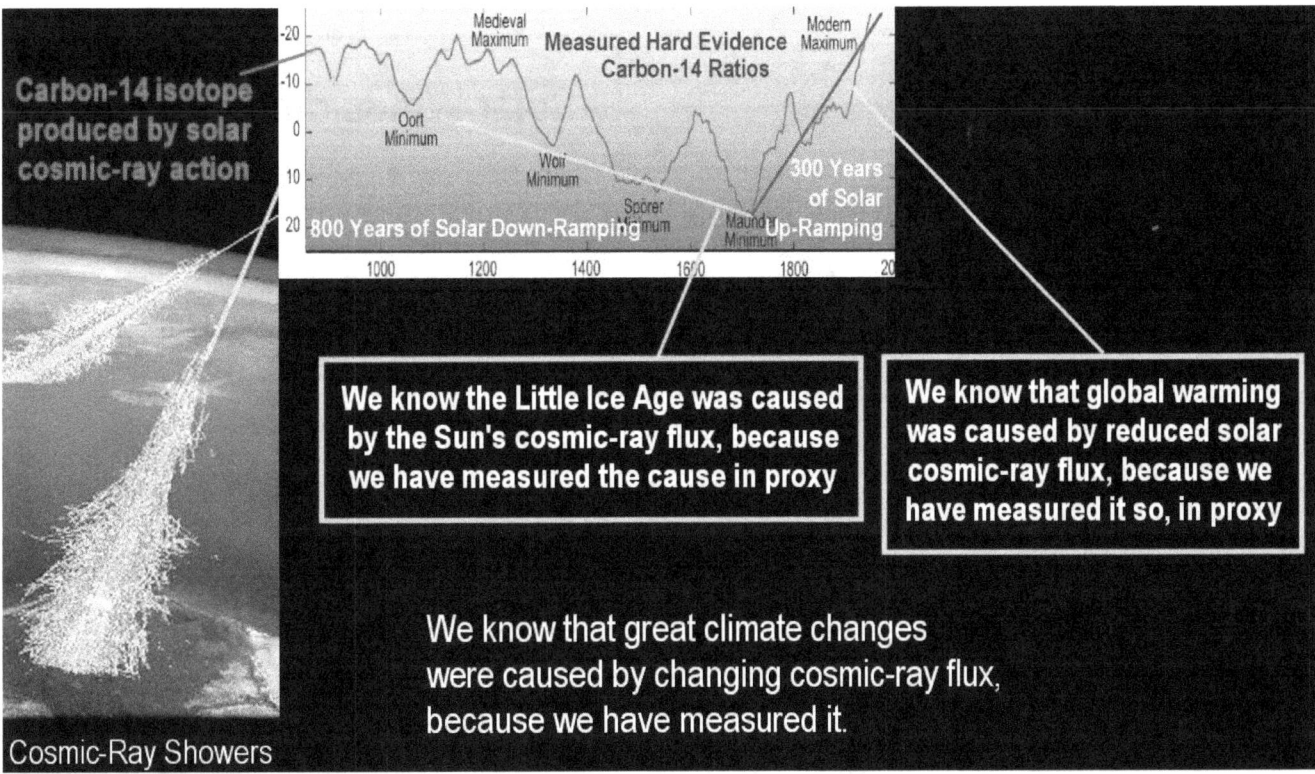

So, how do we know all this? We know this, because we have measured it. We know that great climate changes were caused by changing solar cosmic-ray flux, because we have measured the cause for the changes in ratios of carbon-14 that is produced exclusively by cosmic-ray flux. We have measured the cause for the Little Ice Age that way, and for the global warming that followed. It all comes down to one point: changing solar cosmic-ray flux.

Changing solar cosmic-ray flux is the climate driver

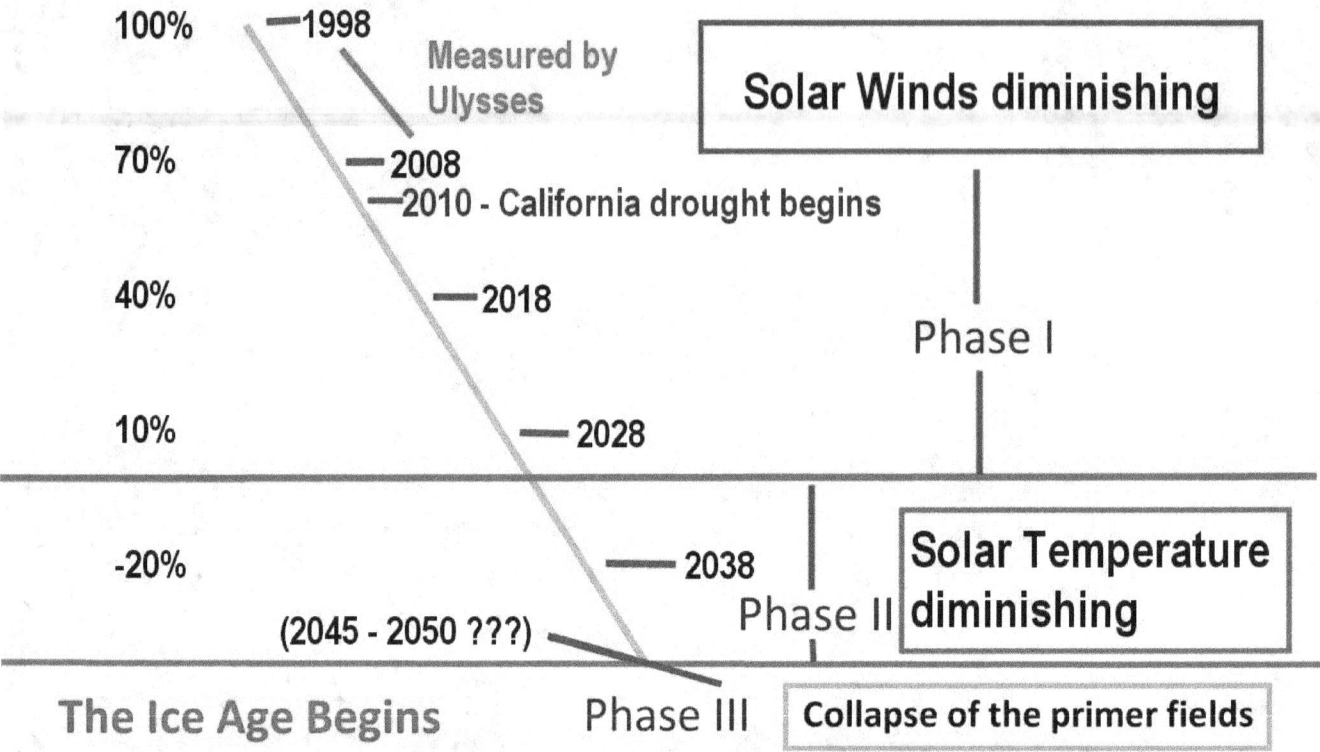

Changing solar cosmic-ray flux is the climate driver throughout the entire Phase-1 of the boundary zone that is now upon us. While the diminishing solar-wind pressure delineates this first phase of the weakening of the Sun, the solar wind itself isn't a causative factor, but is an indicator of how far the system has collapsed that powers our Sun.

Throughout Phase-1 of the solar weakening, the Sun maintains its surface temperature of 5,778 degrees Kelvin. Only the cosmic-ray flux increases during the Phase-1 period, and its climate effects increase with it.

All of the dramatic weather phenomena that have been observed in the beginning of this phase are now becoming more prominent as we have reached the half-way point of this phase of the weakening solar system, with 15 more years yet to come, with the Earth getting colder, years after year.

Every year the Earth gets colder

Climate cooling is accelerating

Every year the Earth gets colder, increasingly

Climate cooling is accelerating. Every year the Earth gets colder, increasingly.

The solar cosmic-ray increase that Ulysses had measured

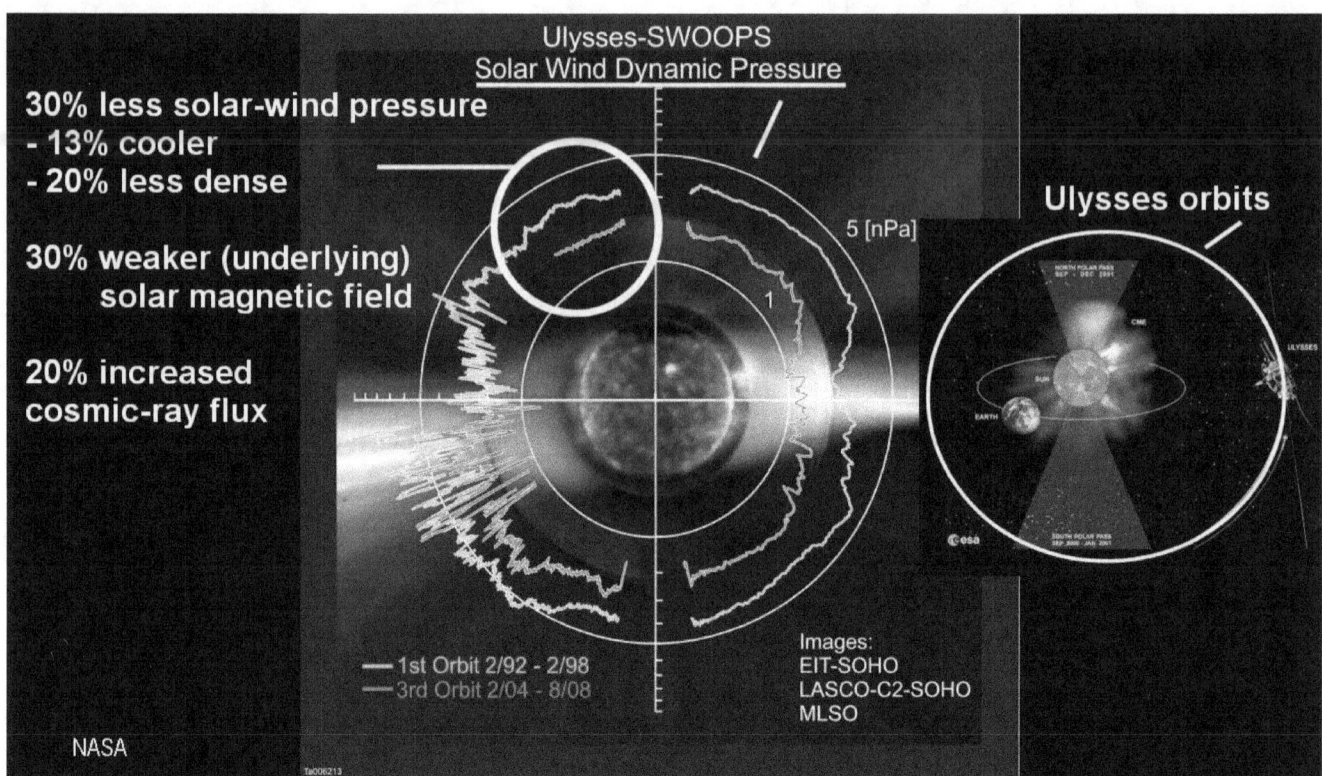

The solar cosmic-ray increase of 20% that Ulysses had measured between 1996-2008, appears to have grown much larger thereafter, according to more recent measurements that measure the effect of cosmic-ray flux in the atmosphere.

Cosmic-ray interactions release of free neutrons

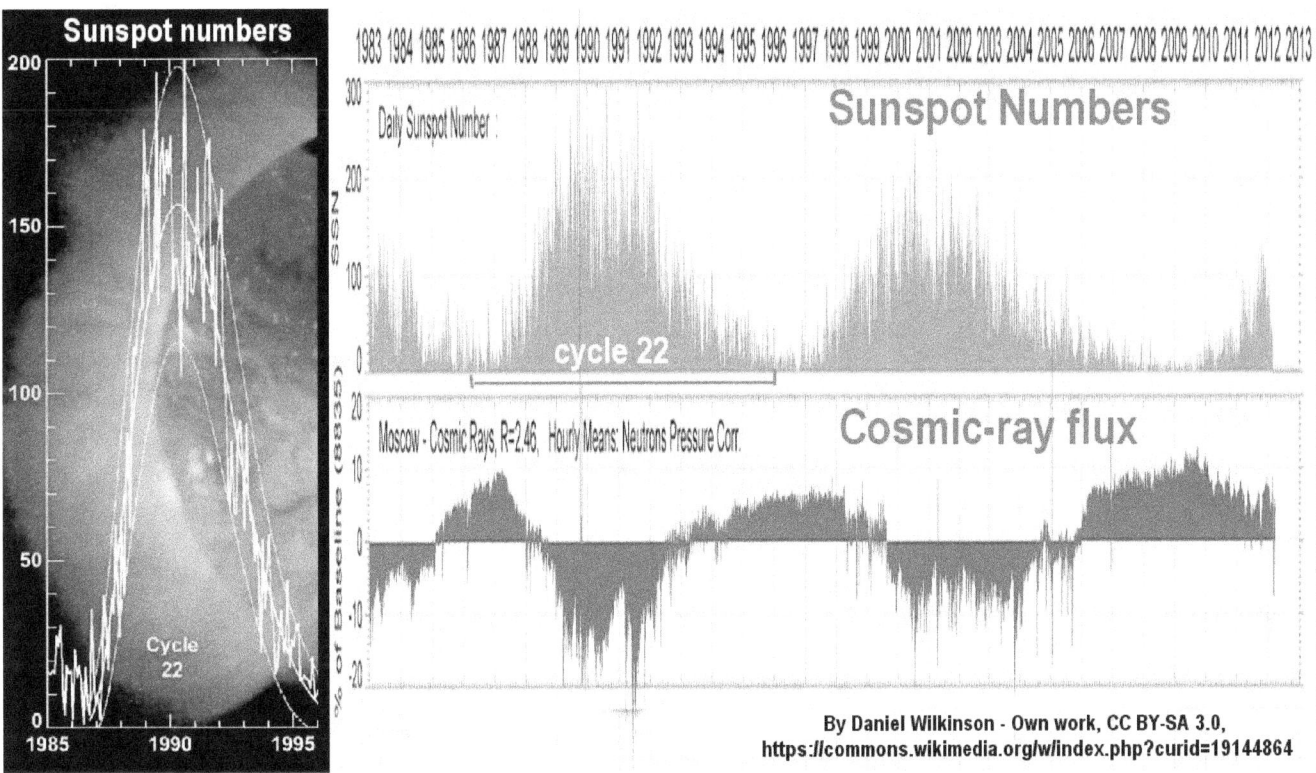

By Daniel Wilkinson - Own work, CC BY-SA 3.0,
https://commons.wikimedia.org/w/index.php?curid=19144864

In considering that cosmic-ray interactions in the atmosphere causes the release of free neutrons, the measurement of atmospheric neutrons provides a modern real-time proxy for changing cosmic-ray flux. The measurements that are conducted on this basis, indicate that the cosmic-ray flux had nearly doubled between the start of solar cycle 23, and the start of solar cycle 24.

It may well be possible the the cosmic-ray flux may double again, even before the end of solar cycle 24, that's not shown here. Should this doubling happen, which is highly possible, the climate consequences on Earth would be severely increased.

If one takes a closer look at the Phase-1 period

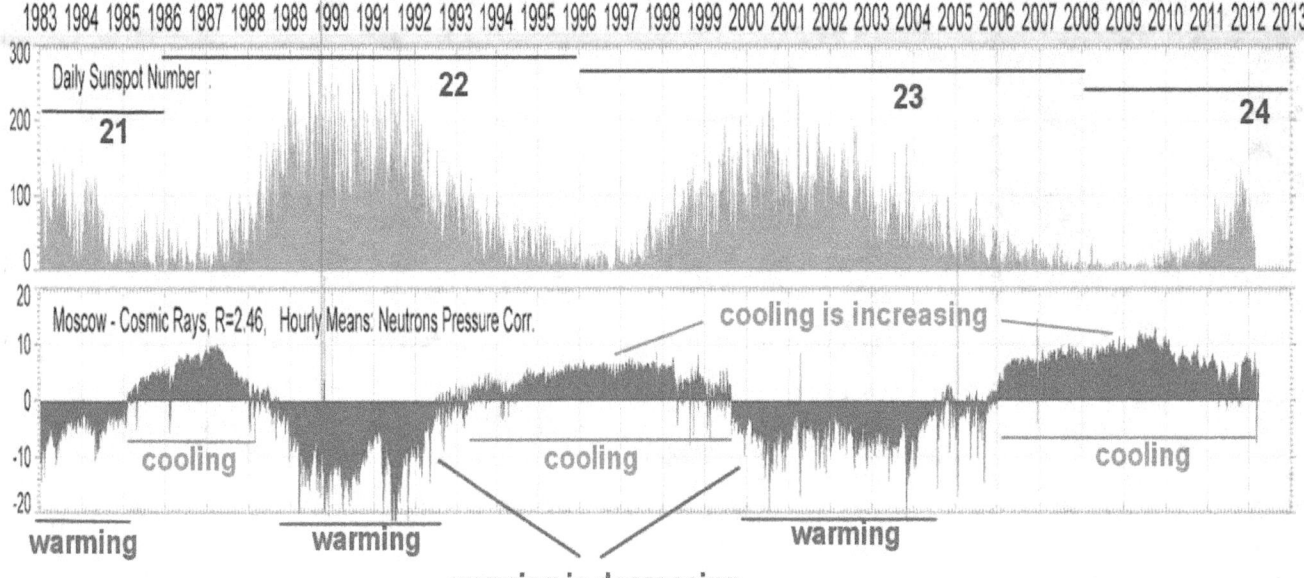

If one takes a closer look at the Phase-1 period, in terms of measured cosmic-ray flux, it becomes apparent that during the peak portion of the 'big' solar cycles, the cosmic-ray flux drops to low values. These low-flux periods, are effectively periods of climate warming. The progression that we see here, between solar cycles 22 and 23, indicate that the warming periods became smaller in amplitude, while the amplitude of the cooling periods was increasing. This means the cooling of the climate is increasing.

It also becomes apparent that the warming periods require a high level of solar activity, for them to happen, and to be maintained. For solar cycle 22, this condition was met 2 1/2 years from the start of the cycle. For cycle 23, which is a weaker cycle, the warming started 3 1/2 years into the cycle. But for cycle 24, for the first 4 years that we have measurements of, which includes the first peak in sunspot numbers of cycle 24, the warming that should have happened, didn't happen at all. The cosmic-ray values remained high all the way through to the end of the graph.

It appears that cycle 24 may be already too weak for the breakout from cooling to warming to happen. This means that we appear to have crossed the threshold already past which the re-warming of the climate during the high portion of the solar cycles no longer happens, because the system that drives the cycles has weakened too far.

Record-high level of the cosmic-ray flux in 2018

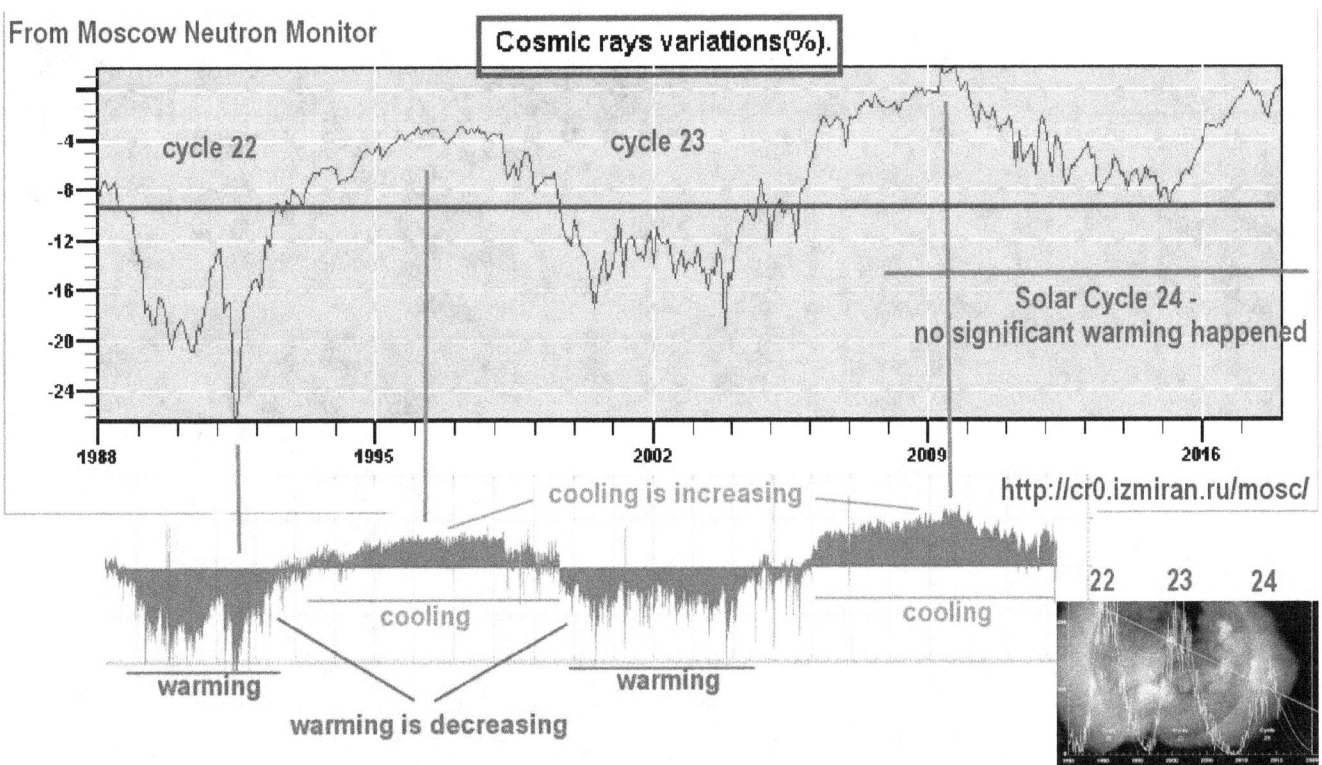

If one extends the graph for the real-time proxy for solar cosmic-ray flux, all the way to the present, it becomes apparent that we have now passed the point where the peaks of future solar cycles correspond with periods of climate warming. While the cosmic-ray flux for the current cycle 24 diminished slightly during the peak of the cycle, its level remained high throughout the peak of the cycle, and then increased again to record levels.

It is this record-high level of the cosmic-ray flux that we now experience in 2018. That's what is causing the extremely cold climate anomalies that we now experiencing in many parts of the world, like winter blizzards in spring, snowfalls in June, and Russian tankers being trapped by sea-ice in July.

If cycle 24 doesn't make the grade for a climate re-warming

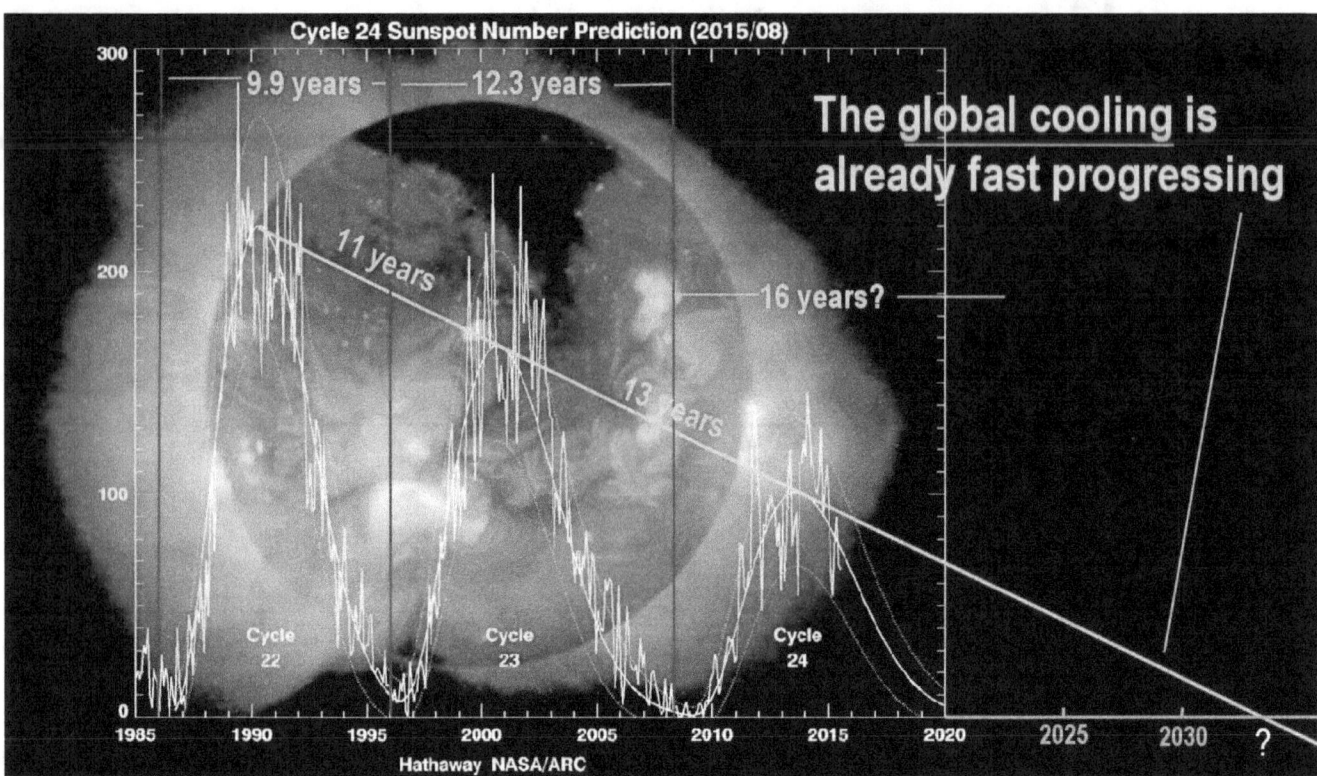

This means further, that if cycle 24 doesn't make the grade for a climate re-warming to happen, none of the following cycles, which promise to be progressively weaker, won't make the grade either. We face a doubly critical situation here, in which increased cloudiness keeps on increasing and the Earth grows ever colder at an increasing rate, since the traditional mid-cycle re-warming by periods of reduced cloudiness, no longer happens.

When solar energy is reflected back into space

When solar energy is reflected back into space by the white top of clouds, the reflected energy is lost to the Earth's thermal budget. The loss to the budget then begins to accumulate if it isn't replenished periodically. As the result, the Earth's climate is getting progressively colder, and as of recently at an ever-faster rate. That's where we are today, without a recourse in sight.

This should raise some eyebrows, because the consequences are scary.

The scene that you see here is scary

The scene that you see here is scary. It is a winter blizzard scene in the middle of April in Nebraska. The scene is of Winter Storm Xanto that brought record snowfall and cold across the North American grain belt. This winter blizzard occurring in spring, by the name Xanto, was declared to have been the second-heaviest snowstorm of all-time in Green Bay, Wisconsin.

It is scary to realize that this late winter blizzard was not a freak anomaly, but was merely the most recent effect of a potentially 15-year trend of the Earth getting rapidly colder, year after year, after which it gets still worse.

Evermore strange-looking cloud formations

The continuing cosmic-ray increase may be the reason why we are now seeing evermore strange-looking cloud formations and experience unusual climate conditions.

Pilgrims freezing to death in Morocco

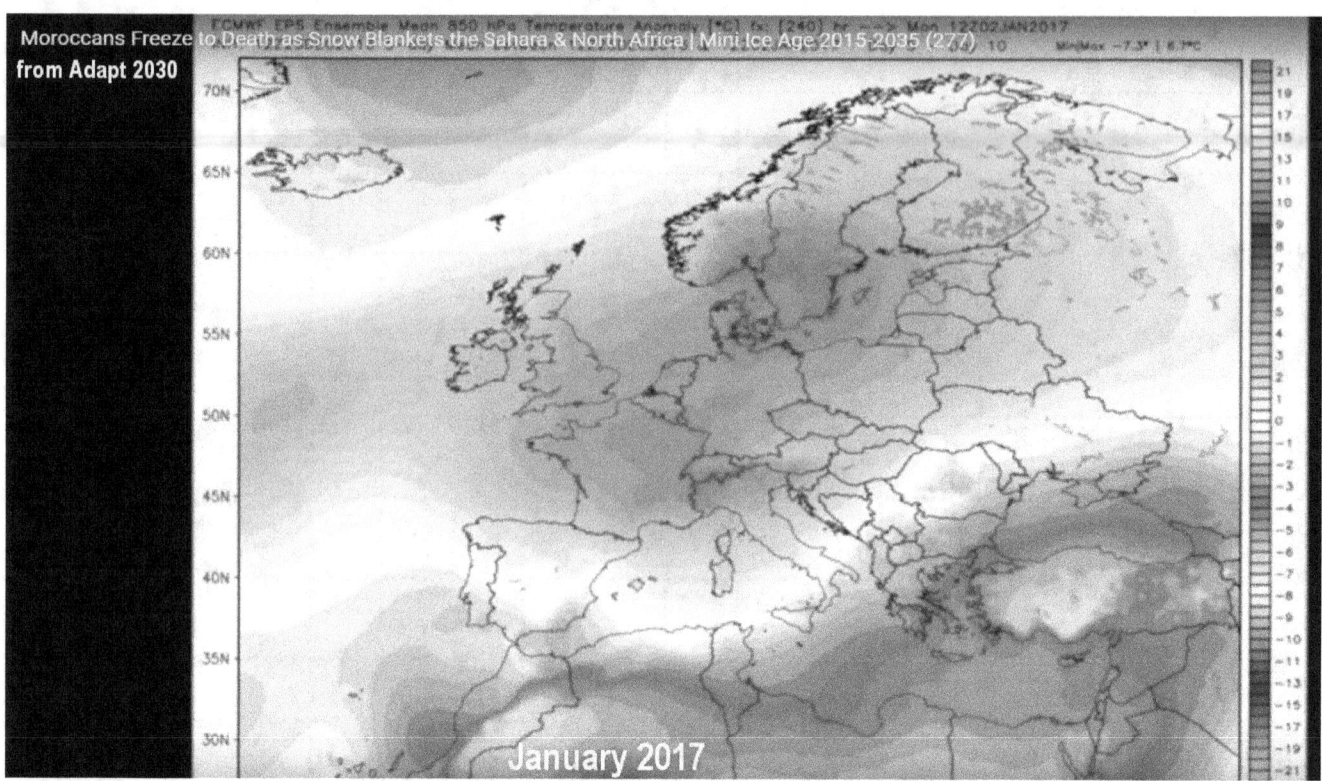

An example is the intense cold-snap across the Sahara in January 2017, which resulted in a number of pilgrims freezing to death in Morocco, who were tracking through the mountains clothed for the traditional hot climate, but were not prepared for the deep-below-zero conditions that surprised them.

Australian wheat crop suffered severe drought

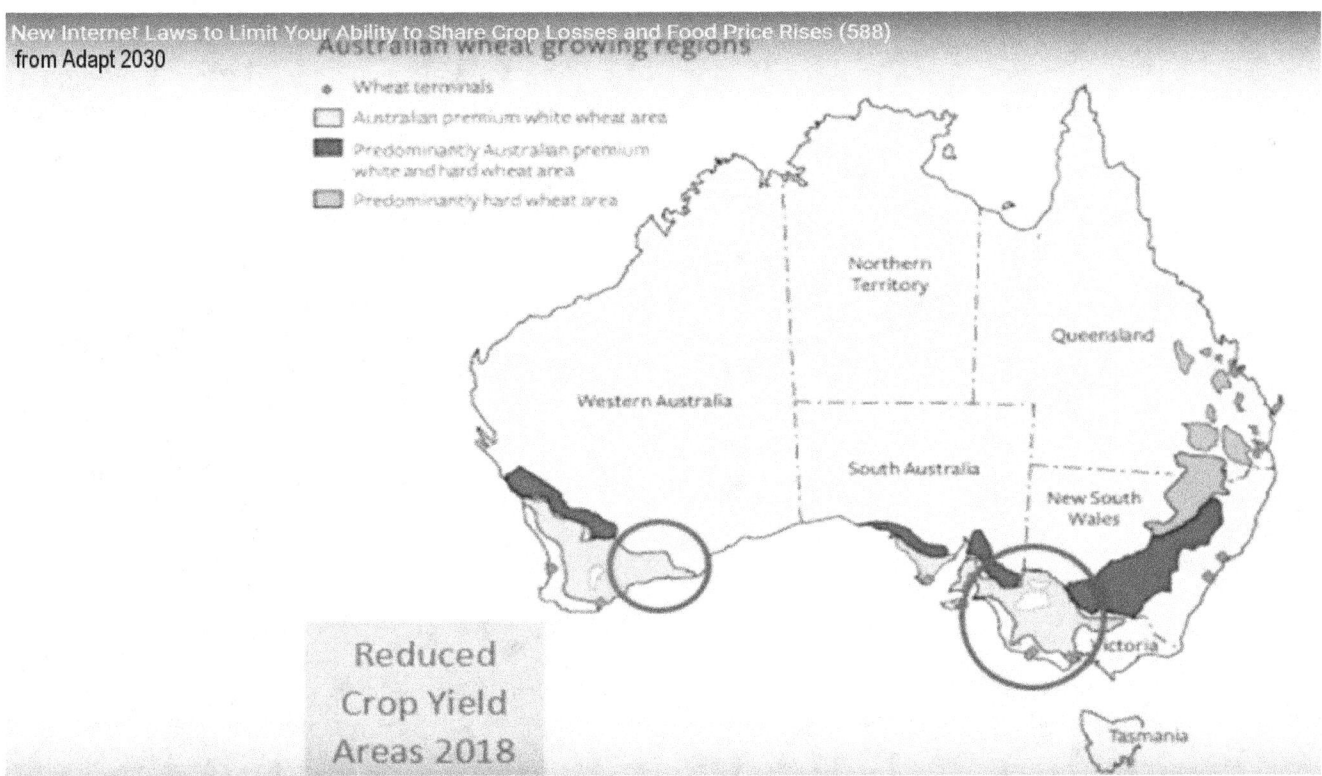

Later in the same year in which the pilgrims froze to death in northern Africa, the Australian wheat crop suffered severe drought and too much heat, so that the 2018 national harvest came in 30% short.

Inversely, during the years of lesser volumes of cosmic-ray flux

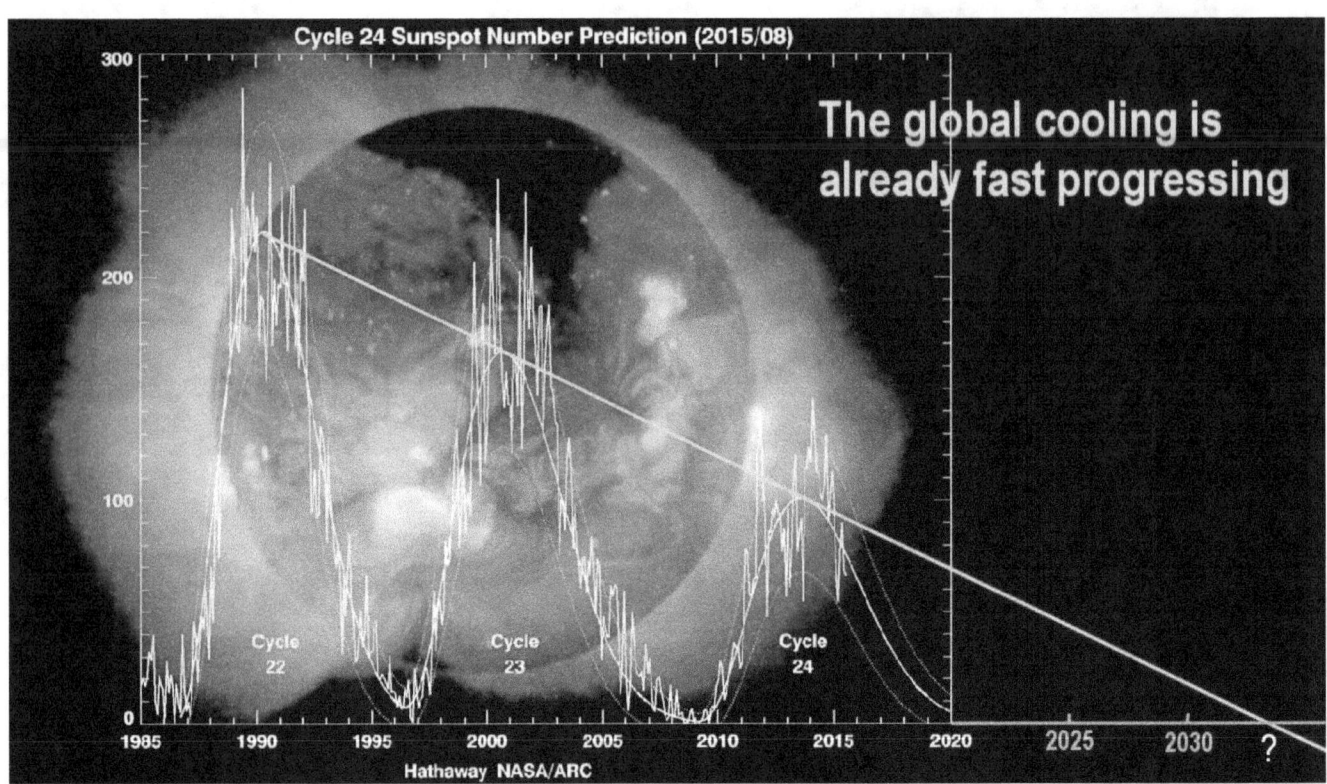

Inversely, during the years of lesser volumes of cosmic-ray flux, during the solar maximum of cycle 24, the Canadian wheat farmers had harvested bumper crops in 2013, 2014, going into 2016, because of 'good' and moist climate conditions in these years.

What if the growing window shortens further?

But now in 2018, as the solar cycle 24 is 'fading' and larger volumes of cosmic-ray flux are impacting the Earth, so much so that the giant April blizzard blew across the North American grain belt, the resulting snow and freezing temperatures had delayed planting that might result in reduced harvests in a few localized areas.

In most cases the remaining growing season will likely be still sufficient to cover the needed 110 to 130 days growing period that wheat crops require. The delayed planting places the harvest time deep into September, where early frost becomes a danger. With a bit of luck the frosts may not happen, and all will work out this time.

But what if the growing window shortens further? The growing window typically opens 6 days after the last frost in spring, and closes 6 days prior to the first autumn freezing. While the current window might still be generally sufficient, it likely has become already marginal in many places. Now what about next year?

Consider that the current trend to colder climates has yet a long way to go, with the solar system getting weaker and weaker, and the Earth getting colder year after year. What will happen when the growing window shrinks below what is needed for crops to grow and ripen? What will happen when this happens over a wide area simultaneously? We will likely reach such threshold conditions in 5 to 15 years.

Climate collapse by the 'dying' Sun

Climate collapse has begun.

---

caused by the 'dying' Sun

Climate collapse has begun, caused by the 'dying' Sun

Crisis dimensions, to the point that entire countries loose their food supply

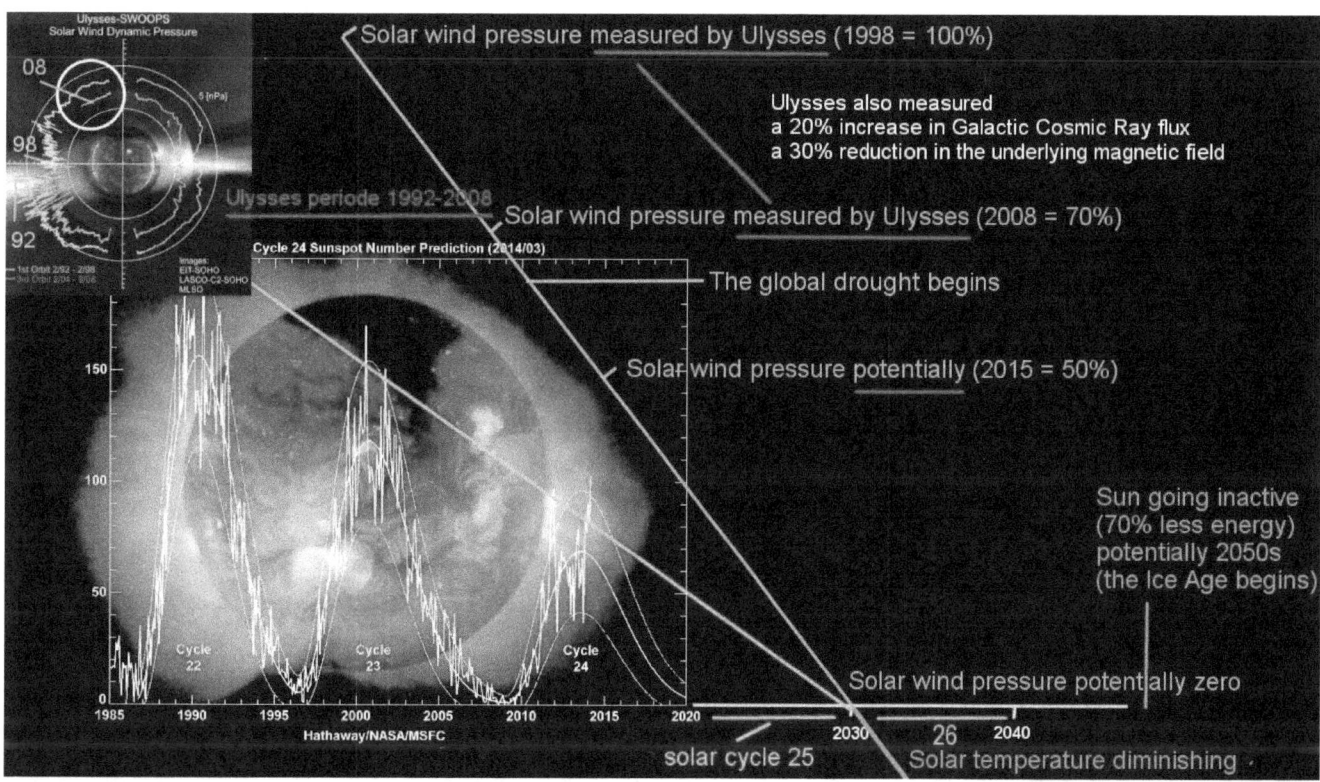

We are facing the potential of a massive climate collapse unfolding over the coming years with a severity that has not been encountered before, except during the Little Ice Age period. Nor will we likely see a recovery from the climate collapse, because the next solar maximum, for cycle 25, that might not occur until the year 2030, if it occurs at all, will likely be so weak that it doesn't affect anything.

The bottom line is that the time of the big solar cycles, like cycle 22, is over, so that potentially harsh times of ever-bigger climate anomalies now lay before us with increasing drought, flooding, storm, and freezing events, all of which deeply affects our agriculture and thereby our food supply. How soon the resulting impact on agriculture will take on crisis dimensions, to the point that entire countries loose their food supply, cannot be determined. The crisis conditions may erupt as soon as 5 years from now, or be as distant as 15 years.

The 15-years point is the point where the first phase will end

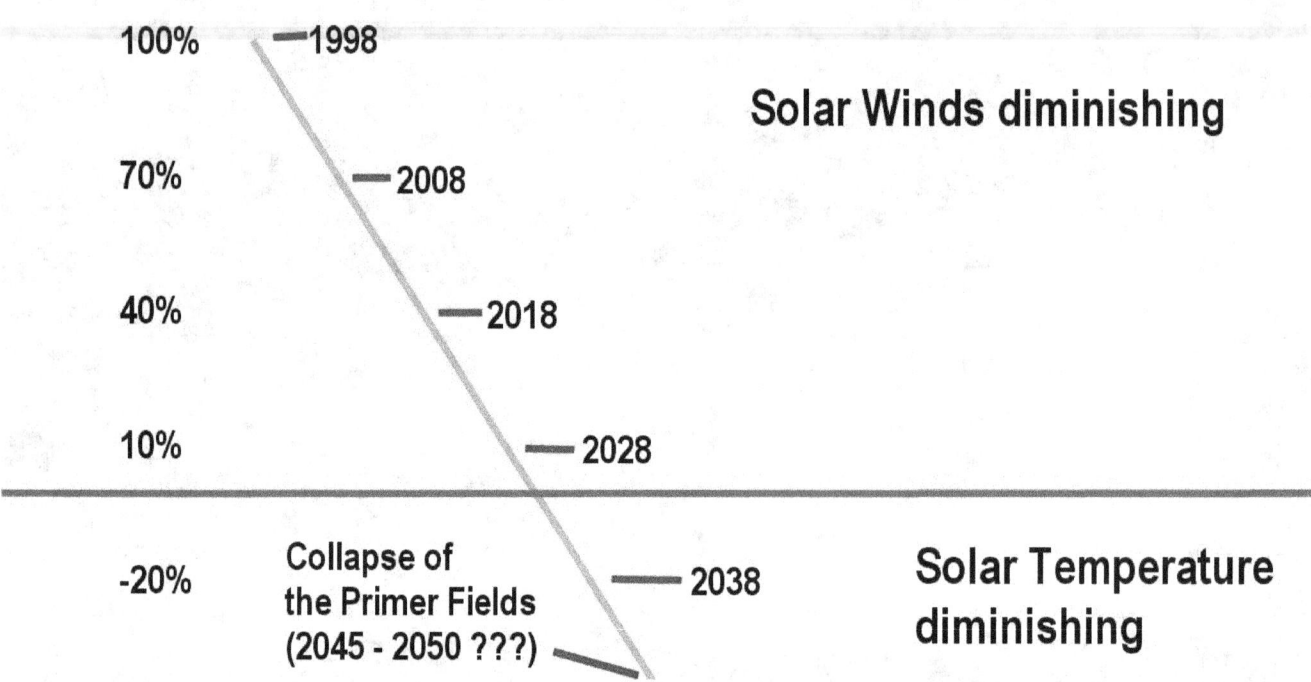

The 15-years point is the point where the first phase of the weakening Sun will end - when the solar-wind pressure will likely have diminished to zero. After that the climate collapse accelerates.

When Phase-2 of the weakening Sun begins

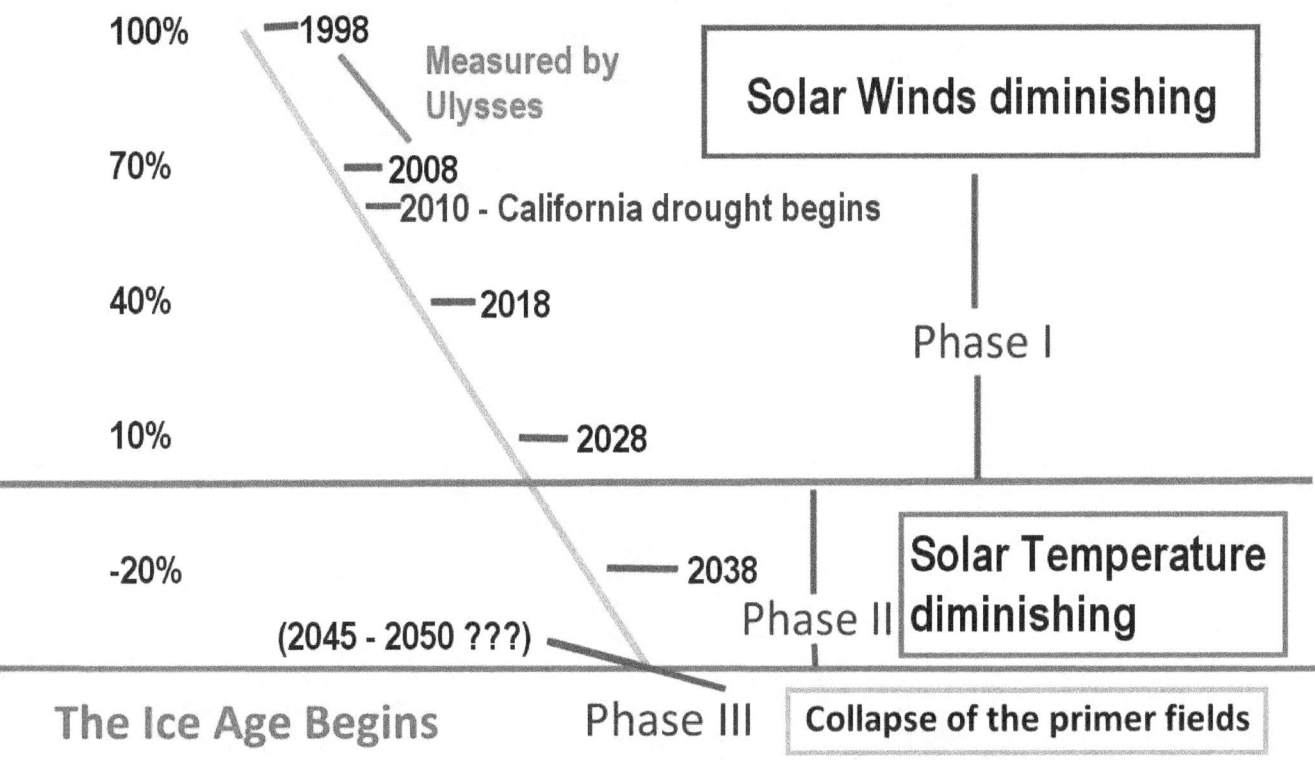

That's when Phase-2 of the weakening Sun begins, in which the solar surface temperature diminishes, and the Sun itself becomes progressively colder. No one can forecast the consequences, nor the actual length of Phase-2, which may be short in duration, or may linger on until the 2050s, when potentially Phase-3 happens.

By the cooling Sun in Phase-2

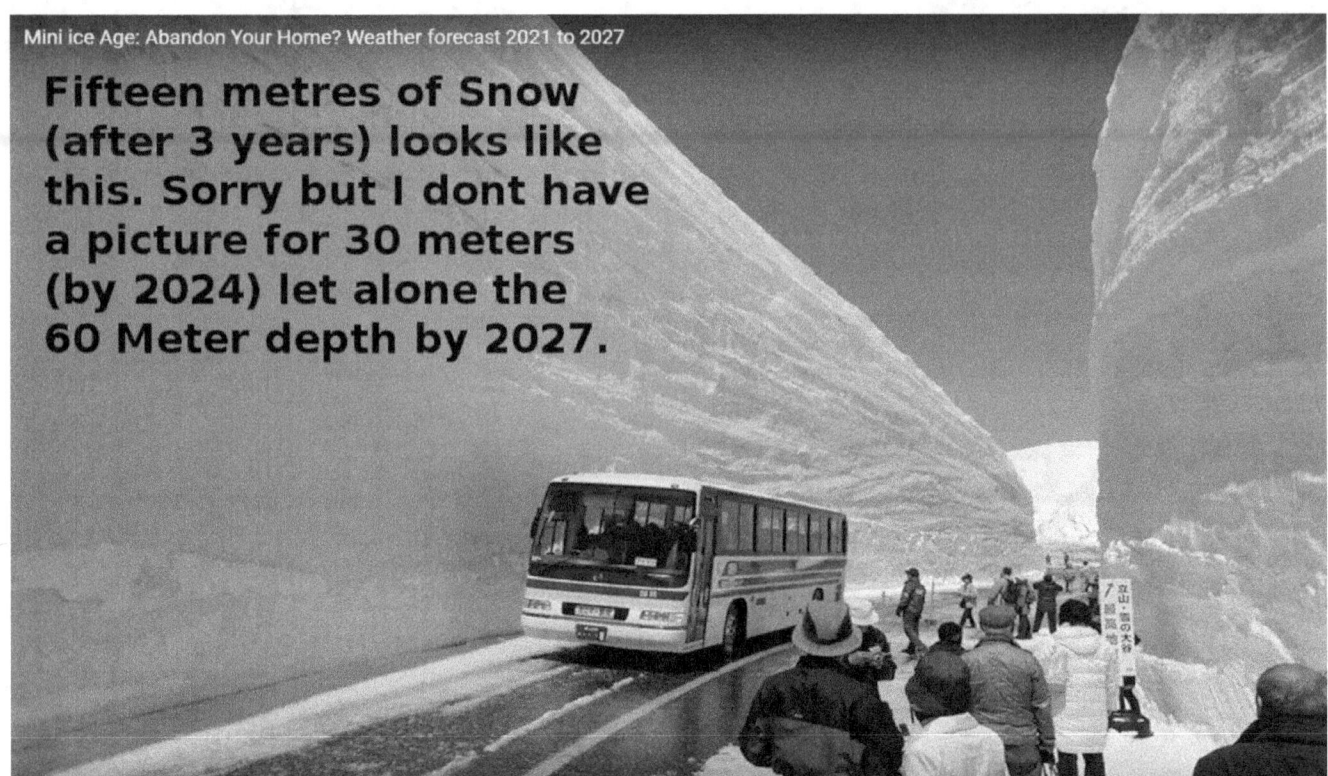

Mini ice Age: Abandon Your Home? Weather forecast 2021 to 2027

Fifteen metres of Snow (after 3 years) looks like this. Sorry but I dont have a picture for 30 meters (by 2024) let alone the 60 Meter depth by 2027.

The accelerated cooling of the Earth by the cooling Sun in Phase-2, will likely become so severe in some of the northern regions that the snow no longer melts in the summer, but piles up and eventually buries towns and cities and breaks down houses and infrastructures. The exact timing, of course cannot be forecast. Such details are left to speculation with a wide margin of uncertainty, but the principle is too imperative for these not to occur. Thus they will occur.

Agriculture will have stopped long ago by then, all through the climate-volatile regions. Agriculture will most likely have collapsed long before the climate transition into Phase-2 happens. When agriculture stops, the greatest human migration will begin, with refugees into the billions streaming into the tropics, fleeing the crippling cold and the resulting hunger, most of them pouring into Africa and South Asia where they would find no refuge as the infrastructures to support their existence would not exist.

The migration of people on such a large scale is unimaginable

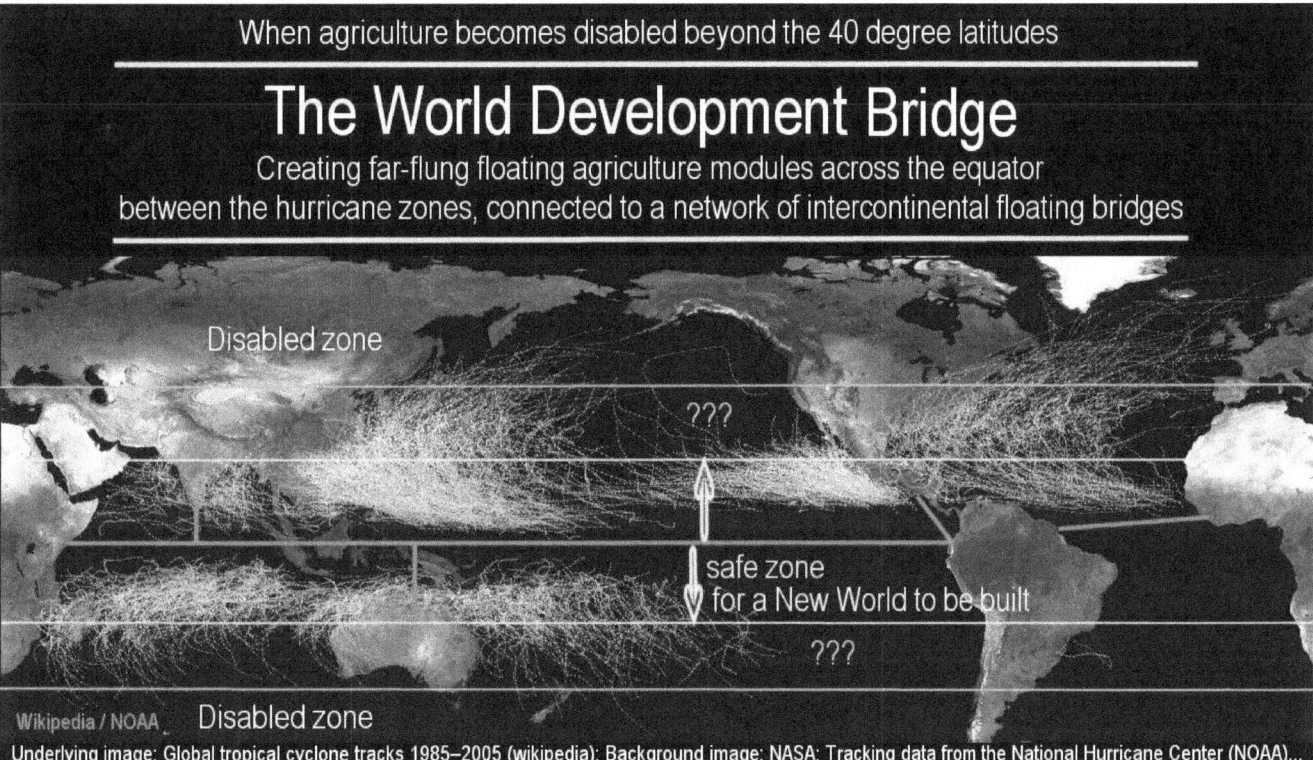

The migration of people on such a large scale is unimaginable, especially in consequences, but it will happen if a new world has not been created by then, by the affected people, which they would already occupy then. The building of the tropical world bridge with floating agricultures and cities along the Equator is the necessary Plan-B for avoiding the unthinkable when Phase-2 unfolds.

The boundary zone ends with Phase-3

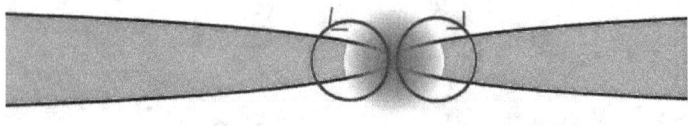

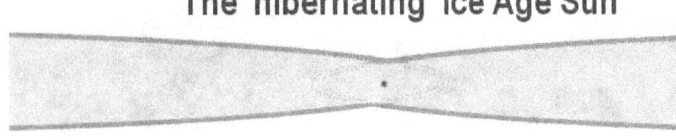

High-density plasma streams with active primer fields operating generate a solar surface temperature of (presently) 5,800 degrees Kelvin

With plasma streams of insufficient density for active primer fields to form, a lower solar surface temperature results, at app. 4,000 degrees Kelvin

The boundary zone ends long thereafter. It ends with Phase-3, when the primer fields collapse that presently focus plasma unto the Sun. When this happens, the Sun goes into a low-powered hibernation mode and the Ice Age begins.

The primer fields are electromagnetic structures that are formed by large volumes of flowing plasma. When the minimal rate of flow that is needed to maintain these structures cannot be achieved, the fields collapse, and their effect - that focuses plasma unto the Sun - vanishes with them.

As a consequence the Sun becomes powered in a default manner by the unfocused plasma stream that then loosely surrounds it.

While the Sun will continue to operate, it will do so with a much-reduced surface temperature, probably in the range of 4,000 degrees, that gives it a 70% reduced energy output that starts the full Ice Age on Earth.

When the full Ice Age is upon us

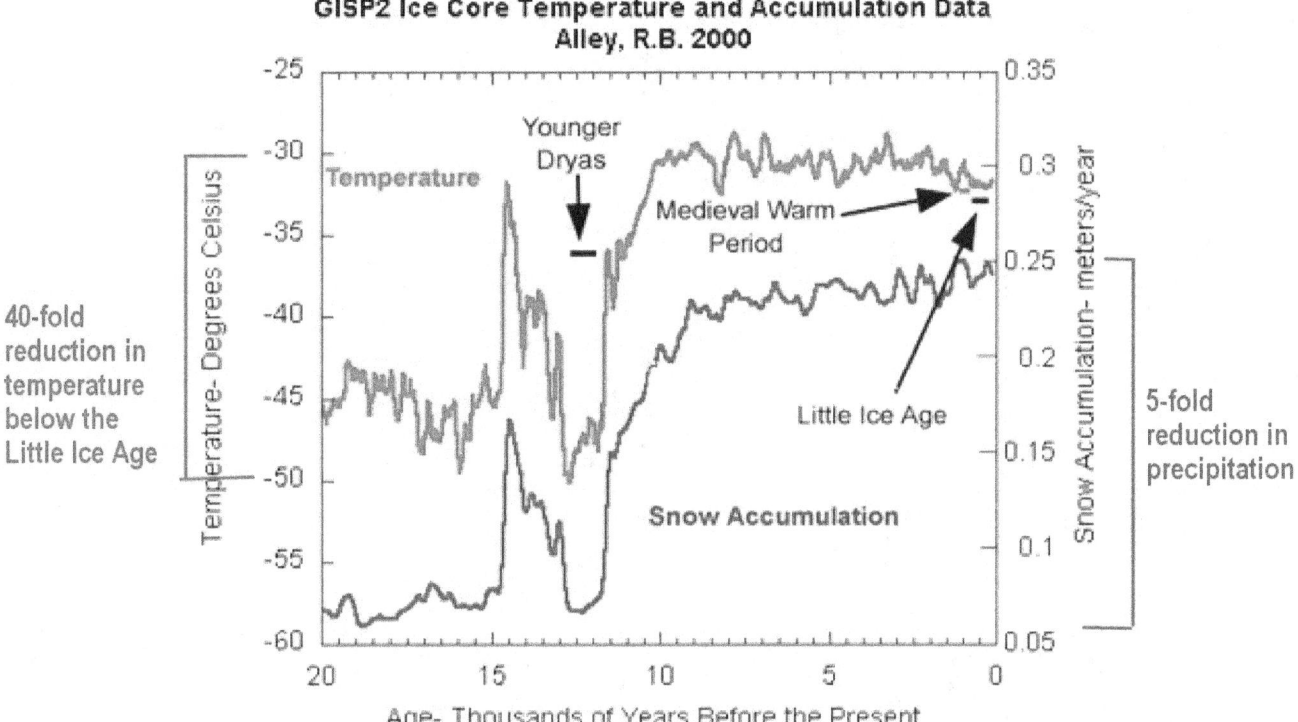

When the full Ice Age is then upon us under the dimmer Sun, the Earth will cool 40 times more extensively than it had cooled during the Little Ice Age. Our planet then becomes transformed into a type of desert ice planet with 80% less precipitation, except in the tropics.

A New World in the tropics as a proverbial Plan-B

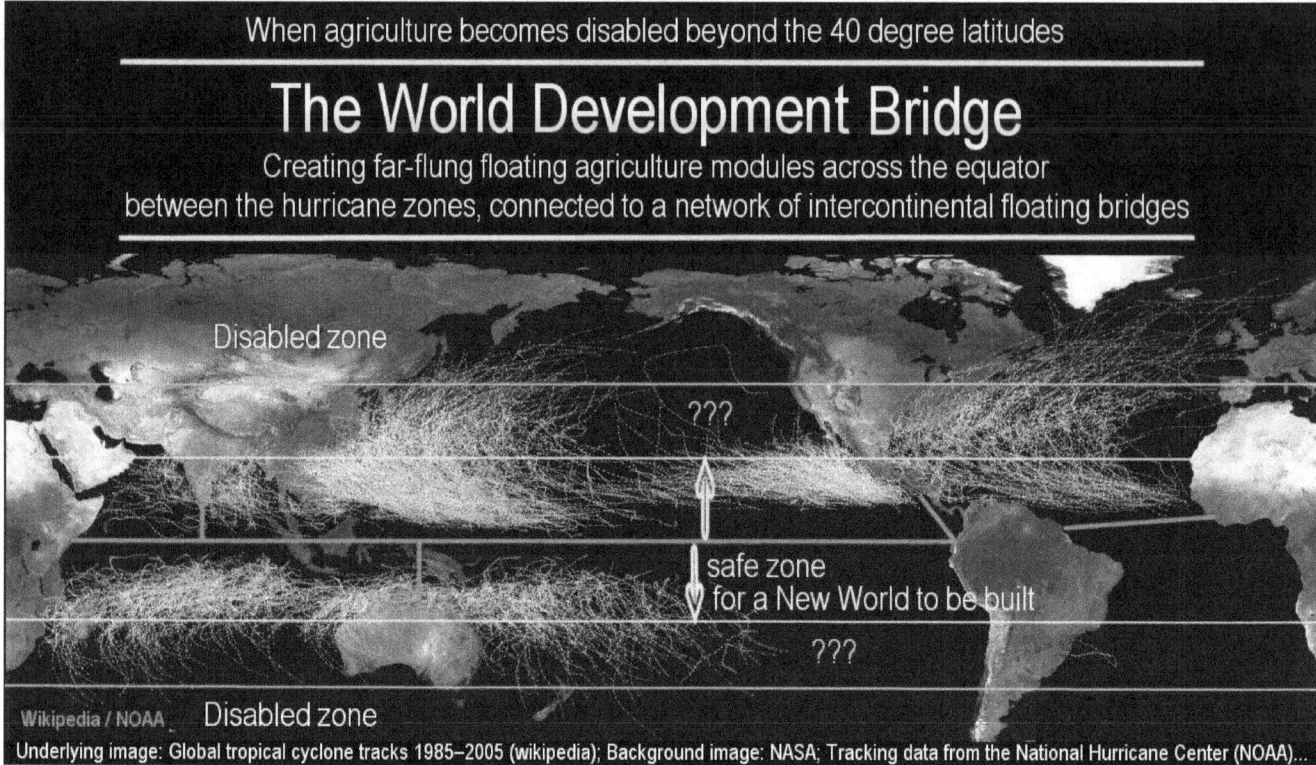

If we have built us a New World in the tropics by then, as a proverbial Plan-B, with technological infrastructures that the Ice Age glaciation cannot affect, then we will continue to live without fail. I would say, to judge by the capable creators that we have become, that this is our destiny - a potentially grand destiny, the grandest ever, grander than what we yet imagine. But will we be true to what we are and reach for it?

Plan-A or Plan-B

**Plan-A
or
Plan-B**

Plan-A or Plan-B

Our plan for survival in the boundary zone to the next Ice Age, is Plan-B. But will we reach for it? The current plan, which is Plan-A, is to do nothing. It's song is, 'let the people die.'

The most critical aspect in the boundary zone to the next Ice Age, that is now upon us, is to choose the right plan, Plan-B, and implemented it quickly and decisively, as time is running out.

The coming Ice Age is assured to happen, because all of the traditional recovery cycles for the Sun have collapsed, and are out of reach. The only long cycle that remains operational is the glacial/interglacial cycle that brings us into the next Ice Age. It is the ultimate cycle of the weakening system that powers the Sun.

Plan-B is our only exit, our only recovery that is possible

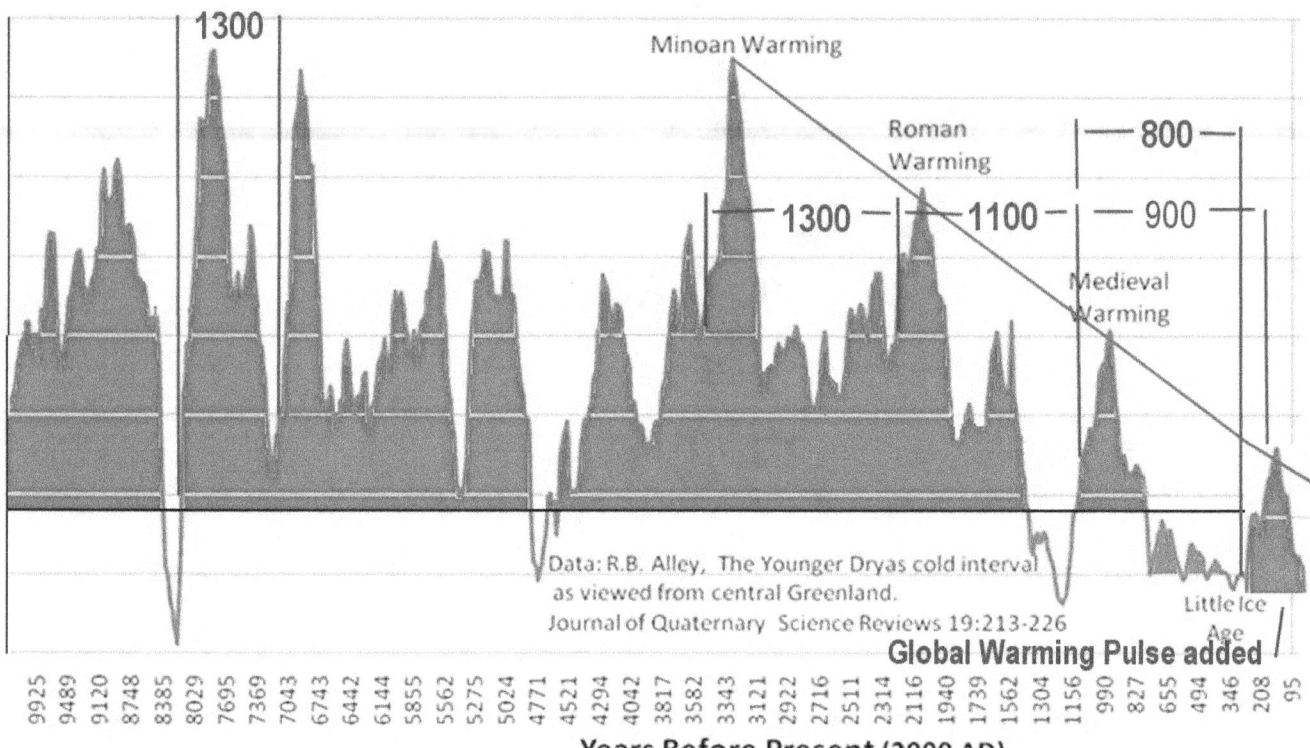

The cooling of the Earth that is now unfolding in the boundary zone is not a cyclical event as we have experienced these in the past. The historic trend of big cyclical warming events has diminished and fizzed out to almost nothing, and what remains is too-far distant to affect us. Even the interval between them has diminished. The global warming cycle that had rescued humanity from the Little Ice Age in the 1600s, had even then, been a relatively small warming cycle. The nest one promises to be too small and still too distant in time, to affect us once again. For all practical purposes the timeframe of the big cyclical warming events has ended. The solar system has become too weak for these cycles to continue. The bottom line is, that there is no cosmic recovery possible out of the boundary zone.

Plan-B is our only exit, our only recovery that is possible, into a created world in which humanity is able to continue to exist and to prosper.

Also the big global cooling cycles have fizzed out

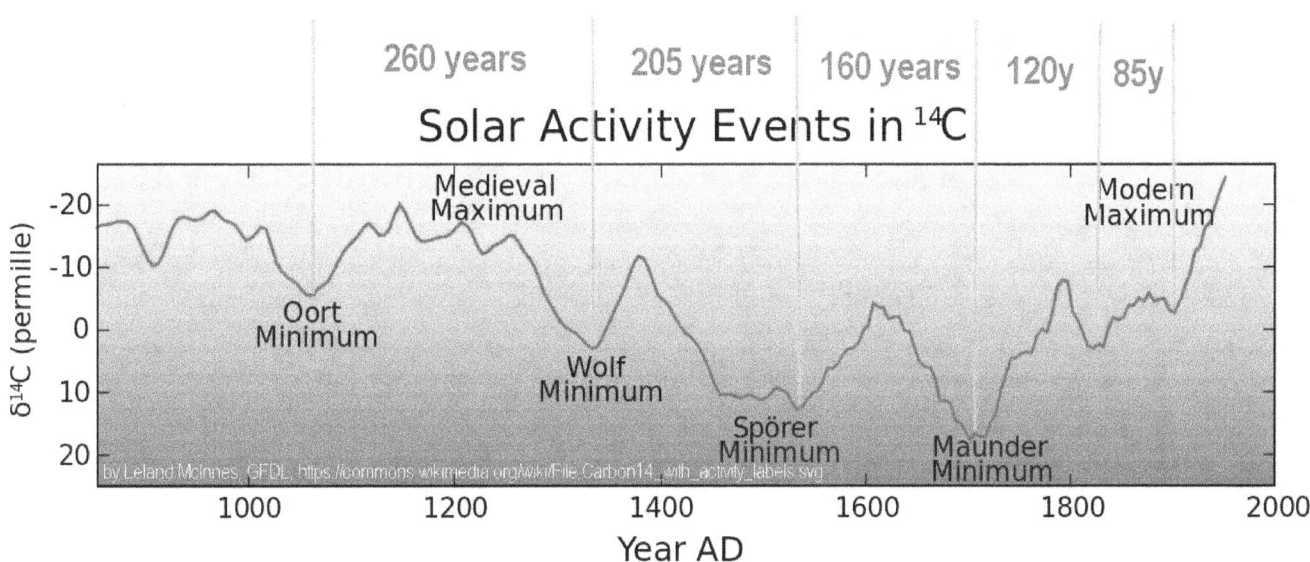

Also the big global cooling cycles that had started 1,000 years ago, have fizzed out into nothing. This means that whoever is dreaming of a "Grand Solar Minimum" followed by a recovery, which is much talked about and anticipated - like a temporary storm that one can weather out - is dreaming indeed.

The historic minimum cycles have all diminished and have fizzed into nothing, and the intervals between them diminished with them.

The intervals have diminished ten-fold

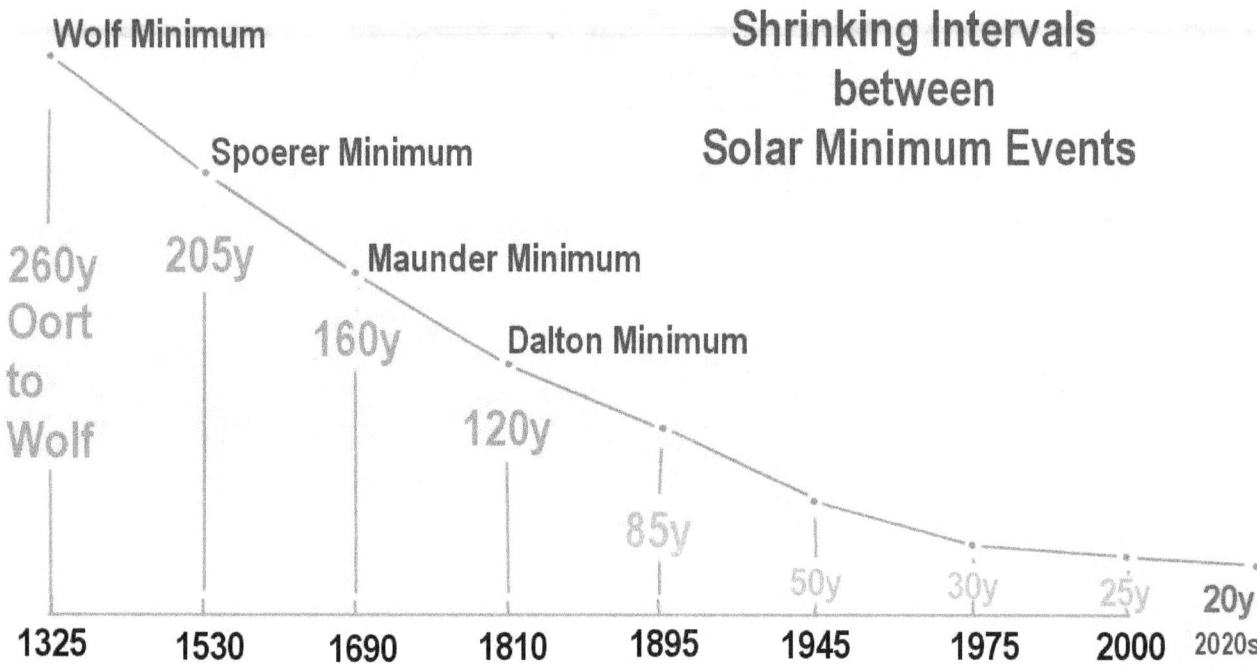

The intervals between these cycles have diminished ten-fold, from 260 years between cycles, down to 25 years. They diminished in an almost geometric progression, and nearly so, as did their amplitudes. The solar system has become too weak for these cyclical phenomena to play a major role again. The time is over when we can predict the future by looking to the past. Who looks to the past is lost in dreaming. This also applies to agriculture in the boundary zone.

The heart-beat slowed from 11 years to 13 years per cycle

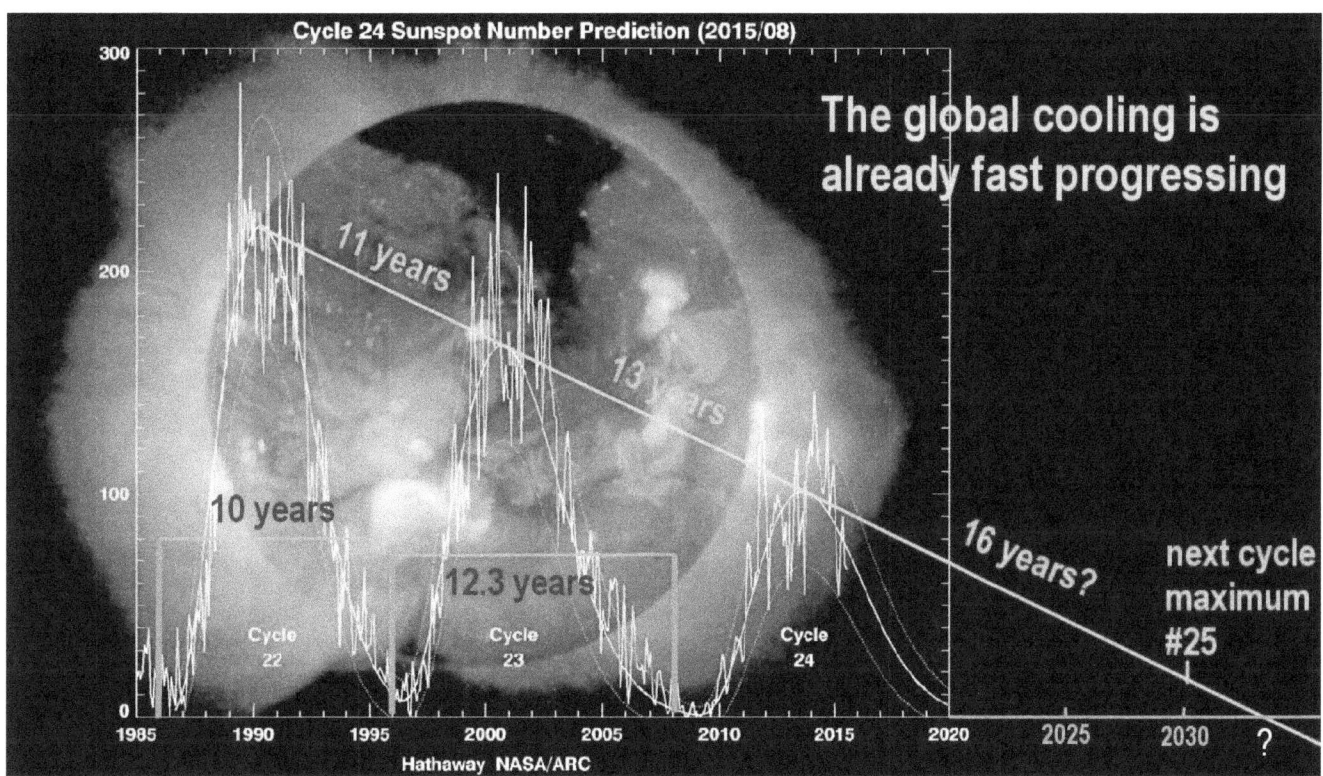

In the boundary zone the collapse of solar activity is evermore accelerating, so much so that the solar system's heart beat, the 11-year solar cycle, is slowing down. The heart-beat slowed from 11 years per cycle to 13 years per cycle, between cycle 23 and 24, and the next one promises to be slower still.

Even the solar cycle itself is beginning to fail

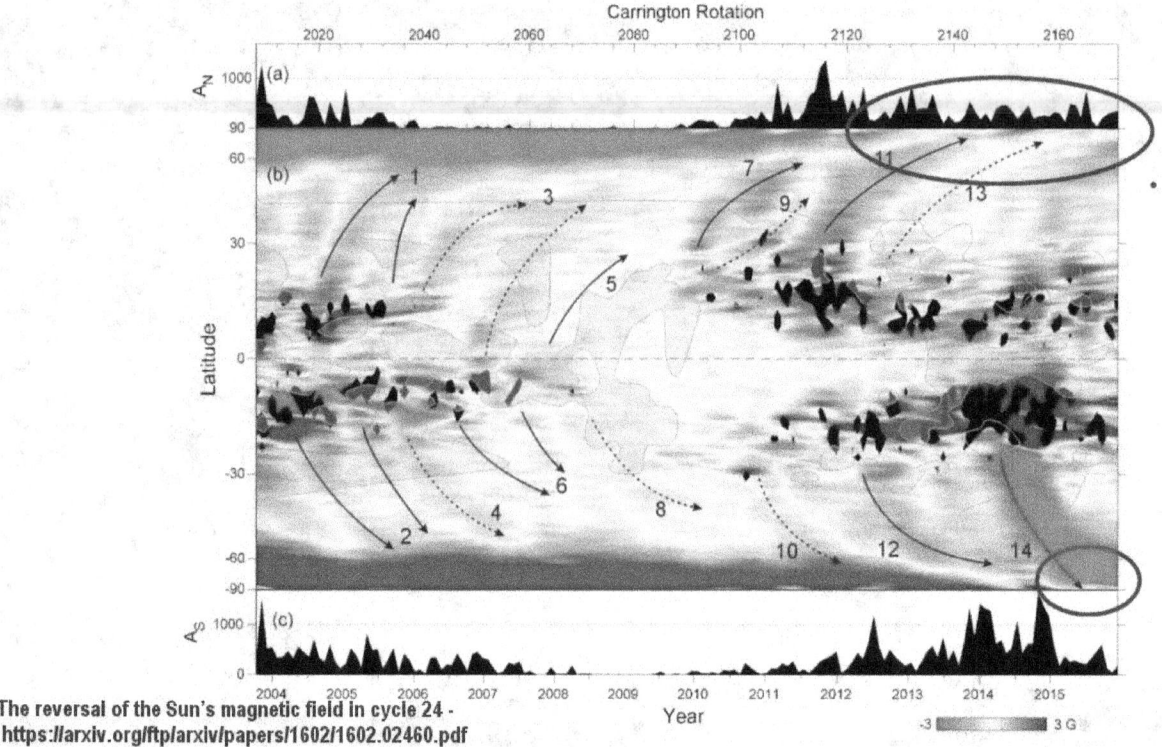

The reversal of the Sun's magnetic field in cycle 24 -
https://arxiv.org/ftp/arxiv/papers/1602/1602.02460.pdf

Even the solar cycle itself is beginning to fail. Normally, at the peak of each solar cycle, the Sun's polar magnetic field orientation flips to the opposite magnetic polarity, but in 2014 only the Sun's South Pole magnetic field flipped, and this faintly, while the North Pole's magnetic field simply fizzed out into nothing.

We won't see anymore decisive recoveries happening

I have explored this fading away of cyclical events extensively in my 3-part mini-series "Grand Solar Minimum becomes the Ice Age." And I have repeated the focus on the cyclical events being history in the video "The Ice Age is Near." We are in the boundary zone now, and as I said before, we won't see anymore decisive recoveries happening, because the solar system has grown to weak for this to be possible. We can only build ourselves out of becoming trapped by the collapsing solar system, by looking forward and build on the scientific knowledge that we have gained of the solar dynamics that create our future climate.

Part 5 - Real-Time Experienced Climate Consequences

Part 5

Real-Time Experienced

_____

Climate Consequences

Part 5 - Real-Time Experienced Climate Consequences

Climate consistent with the changing solar dynamics

We need to look at the climate events that are happening in the present, and see them in the context of what we scientifically know about the weakening of the Sun.

It is not hard to note in this context, that all the climate events that are presently happening, are consistent with the changing solar dynamics and their lawful climate consequences that are experienced all over the world in the form of increased floods, droughts, untimely blizzards, unseasonal frosts, and so forth. Plan-B becomes attractive when we consider the current fringe effects as just the beginning consequences of ever-increasing volumes of solar cosmic-ray flux that escapes our increasingly less-shielded, weakening Sun.

That's Plan-B. That's the only option we have.

With evermore coronal holes opening up in the Sun's atmosphere, cosmic-ray effects on the cloud-forming process on Earth, becomes increasingly unpredictable with large anomalous climate effects occurring. That's dangerous for agriculture, because it affects the available growing season. While we cannot change the solar effects, we do have the option and the power to relocate our agriculture into the tropics where the growing season is more secure, and can be totally secured by technological means. That's Plan-B. That's the only option we have.

Lets look at the case of wheat

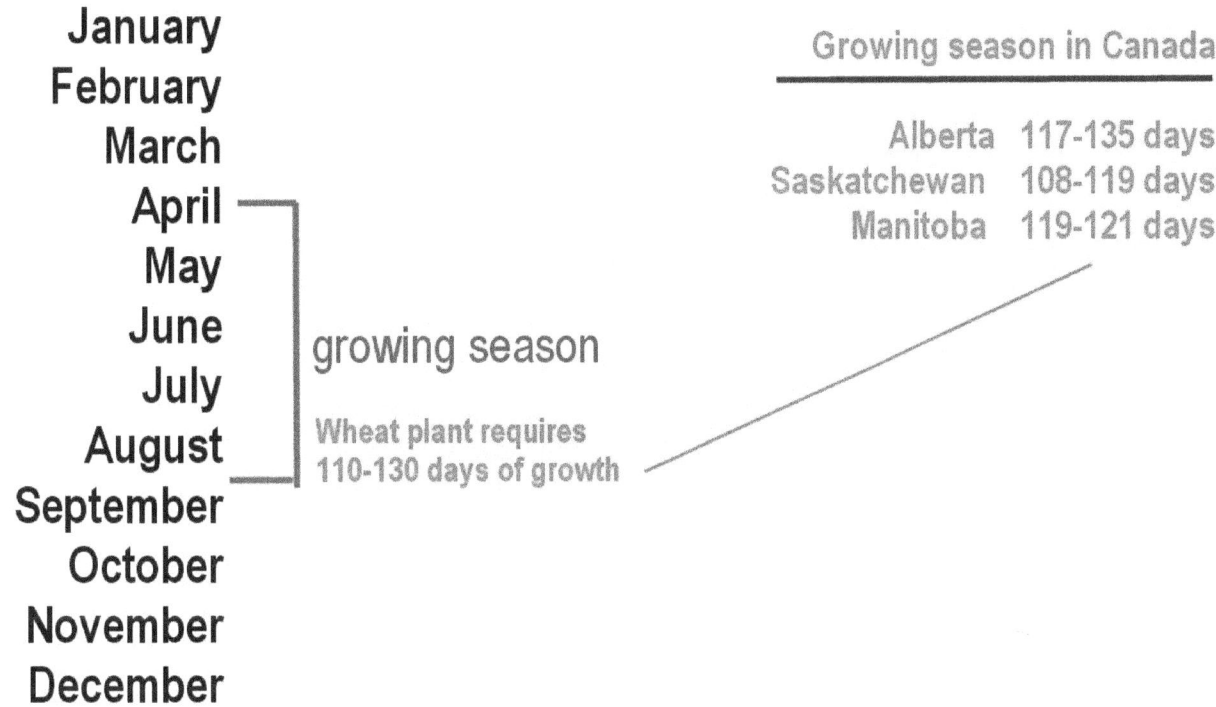

Let's look at the case of wheat. Wheat is one of the most basic foods.

And then, let's look at the case of Canada, which is the world's 6th largest wheat producer, and one of the big exporters. For as long as anyone remembers, wheat has been grown in the three western prairie provinces. Wheat flourishes there because the soil is good and the growing season is sufficiently long, warm, and moist for the plants to mature. Wheat requires 110 to 130 days from planting to harvest. The growing seasons in the Canadian provinces are sufficiently long to meet that requirement, though not with a big margin. If spring is delayed, or interrupted by a blizzard, the required growing window may one day soon not happen. In such cases, when late planting occurs, the plants still grow, but they don't have sufficient growing time to mature. The majoring typically happens in July. As a consequence of weather-delayed planting, the yields are smaller and the quality is poorer, and more so if early frosts or snowfall wreck the harvests.

Winter blizzards in spring

Winter blizzards

in spring

Winter blizzards in spring

The big blizzard, Xanto, struck in April

Is this what happened when the big blizzard, Xanto, struck in April and dumped up to 4 feet of snow in some places, onto the fields across the North American wheat belt, with temperatures dropping low into the freezing range for many days of the month, with Canada being similarly affected.

Some serious discussions about a Plan-B

The outlook is not pleasant. This snow is real. We have come to the end of April and the snow on the fields has not yet melted. How long can the winter wheat remain buried? How soon can the spring planting begin? Will there be enough time remaining? These types of questions should start some serious discussions about a Plan-B.

At the end of April half the planting is typically complete

At the end of April, for example, half the planting is typically complete, but in 2018, nothing was planted across most of the wheat growing regions, and the readiness of the fields for planting was officially considered 'extremely poor.'

No doubt, agriculture will recover from this late blizzard and others like it in the years ahead. Plants will still grow. But, as the 2018 harvest might be an indicator for, future harvests will likely become increasingly "disappointing," as the media like to say.

With a comprehensive Plan-B becoming implemented

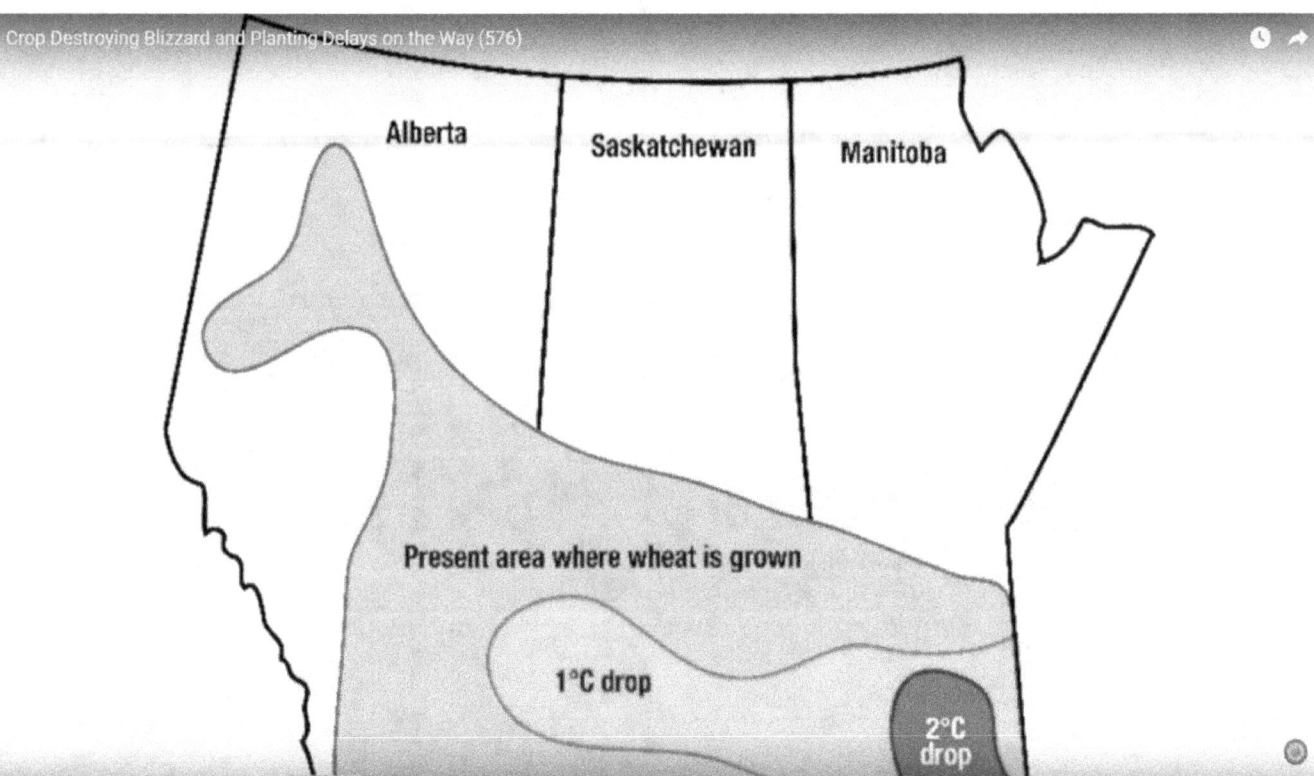

Some studies indicate that an increase in global cooling of just 1 degree Celsius, would shrink the area where wheat can be grown in Canada to less than half, and a further degree of cooling would shrink it to just a tiny bit.

With the current effects being already severe, what will happen when the increasing climate collapse is beginning to affect the world's food production more substantially, and worldwide?

It is hard to imagine how extensively the world might change in the remaining 30 years till the Ice Age phase shift begins, unless humanity intervenes and stages a new direction for its future, with a comprehensive Plan-B becoming implemented.

No one remembered what the Maunder minimum was like

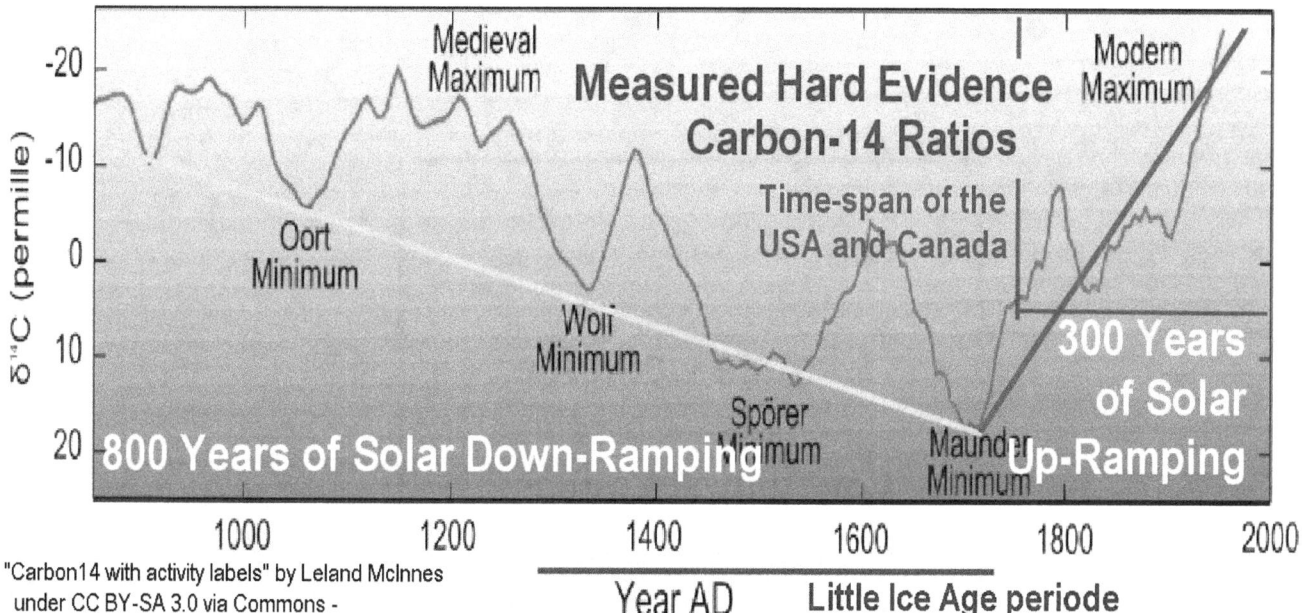

Changing solar cosmic-ray flux, measured in carbon-14 ratios, shows direct inverse relationship with known cold-climate events

"Carbon14 with activity labels" by Leland McInnes under CC BY-SA 3.0 via Commons -

In the early time when the USA and Canada had established themselves as nations, the Sun had already been up-ramped a long way by its last global warming pulse. This means that no one remembered, even back then, what the Maunder minimum was like on the American continent. Nevertheless, that's the climate that will ravish the land again in the not-so-distant future.

Part 6 - Climate Collapse - Agricultural Collapse

Part 6

Climate Collapse

Agricultural Collapse

Part 6 - Climate Collapse - Agricultural Collapse

Covered in permanent snow during the Little Ice Age

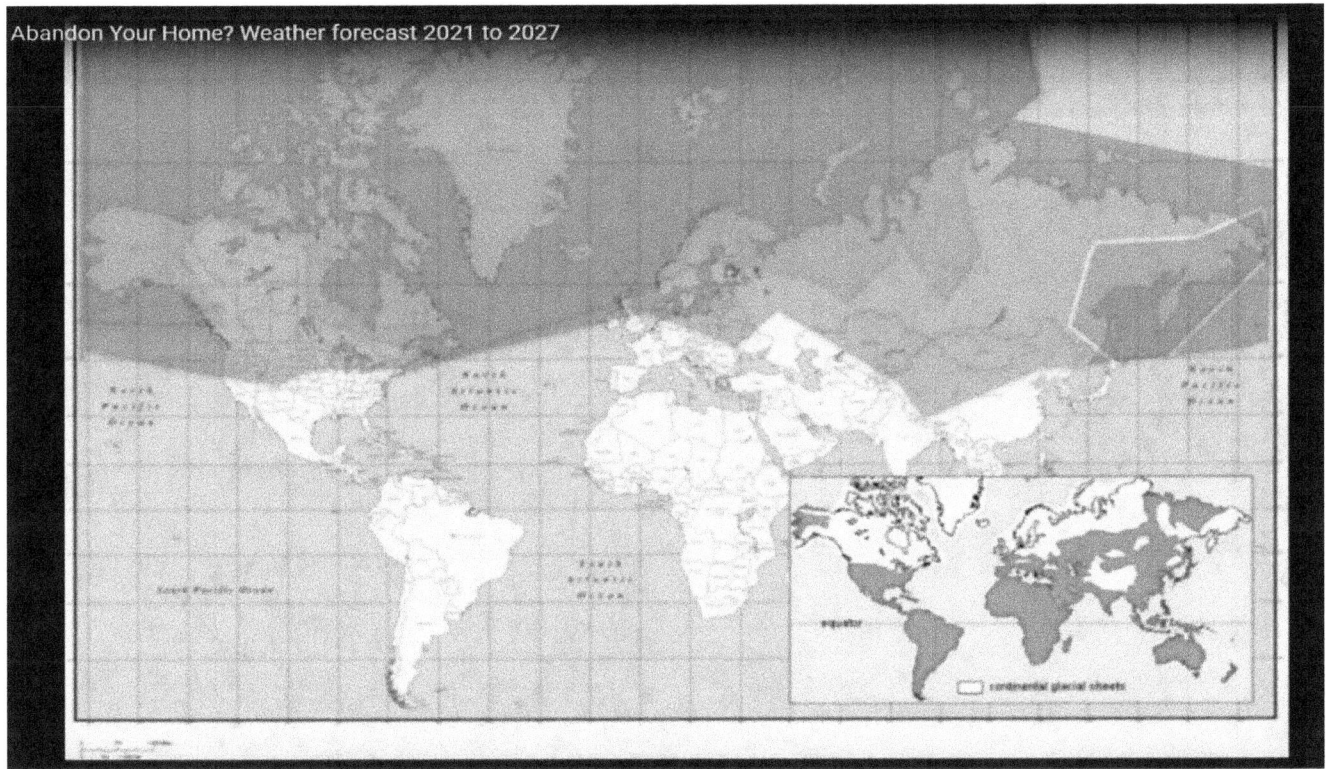

A researcher suggests that most of Canada might have been covered in permanent snow during the Little Ice Age, especially during the Maunder Minimum. A researcher discovered that in northern England and Ireland the snow hadn't melted in the summers in the northern areas during the Little Ice Age, but had remained on the ground and had accumulated. He compared the current climate of this northern region with the current climate around the world.

He mapped the result as two zones. His blue zone marks the area where the snow would have likewise accumulated all-year round. His red zone marks the area where extremely heavy accumulations would likely have occurred.

The cradle for the big ice sheets

His blue zone cuts across the northern USA and encompasses most of Canada. The resulting projection illustrates to some degree where the April blizzard likely had originated that swept deep into the North American grain growing region in 2018.

It appears that to the West of the string of the great lakes, lays the cradle for the big ice sheets that likely carved out the lakes during glaciation conditions.

The outflow would take the cold air mass across the Canadian prairies

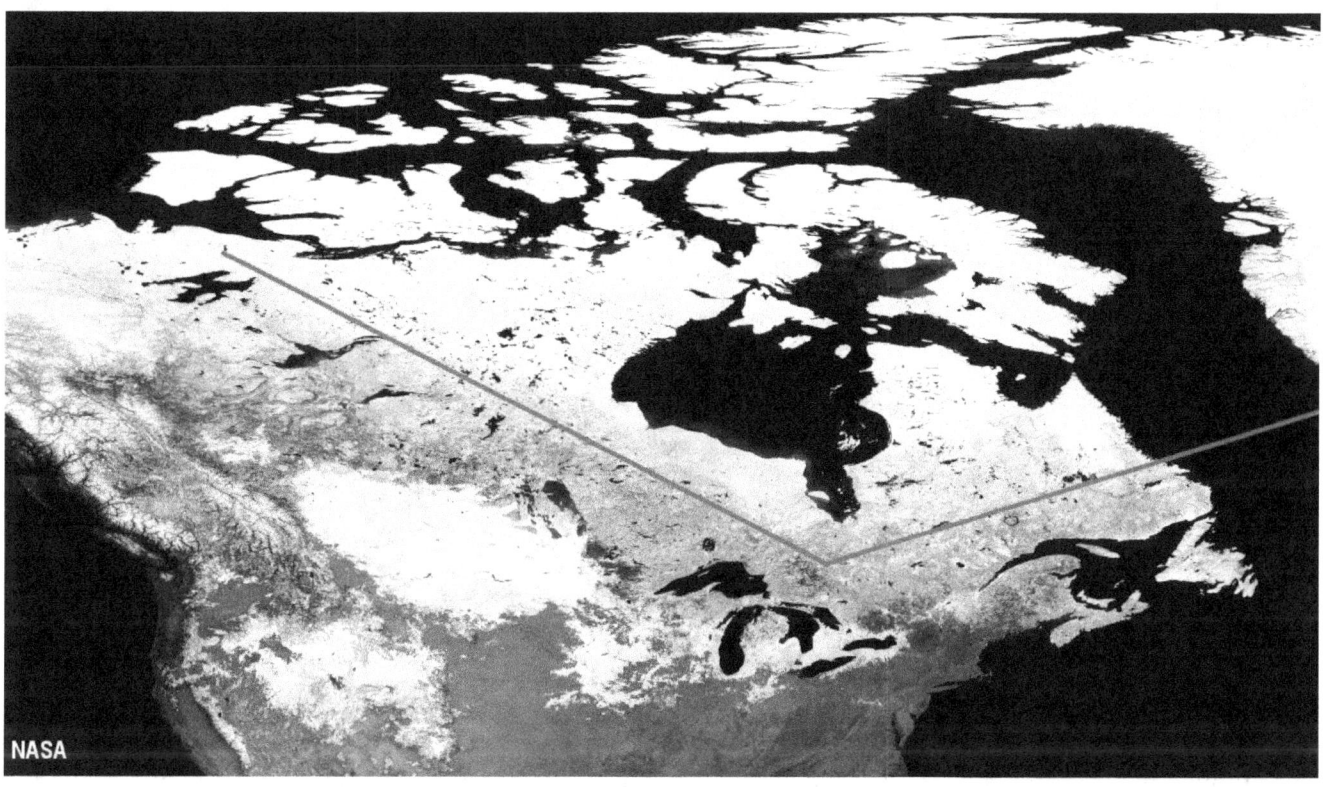

Since latency causes the atmosphere of our planet to flow towards the West by the rotation of the Earth, the moving air mass gets trapped by the Rocky Mountain chain in the West and then becomes forced southward by centrifugal force, the outflow would take the cold air mass across the Canadian prairies into the American grain growing areas.

This means that the longer the large landmass around Hudson Bay and in the North of it remain snow-covered during the year, the greater is the potential for very late blizzards to develop that sweep into the South late in the year, such as happened in 2018 in April.

The big winter-blizzard in spring, named Xanto, of a type for which no official term yet exists, appear to be a new phenomena that will become more common in the unfolding boundary zone, as the Earth is getting colder.

It appears that when the European pilgrims landed in North America in 1620, in the midst of the Little Ice Age, they missed the big cold zone of perpetual snow, although just barely. The pilgrims had landed just south of the cold zone, on Plymous Rock inside Cape Cod Bay.

Though the Pilgrims missed the brunt on the Little Ice Age

The Landing of the Pilgrims, by Henry A. Bacon, 1877

Even though the Pilgrims missed the brunt on the Little Ice Age, their new land was not a picnic ground for them. Plymous is situated 42 degrees, which is the latitude of northern Spain in Europe that had also been gripped by the Little Ice Age.

The Beast from the East

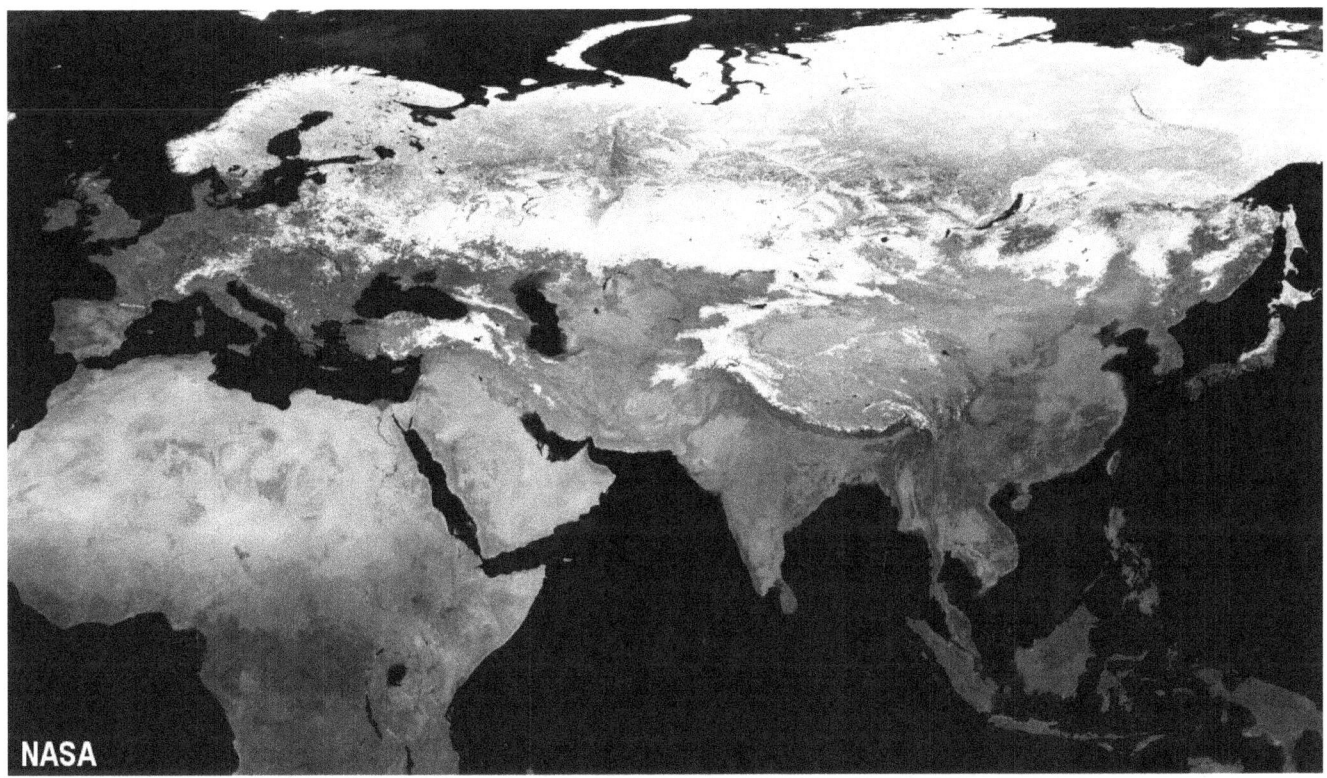

In Europe the cold outflow follows a different route. Because the cold from the large northern landmass tends to flow westward by the latency effect of the rotation of the Earth, and no obstructing mountain ranges stand in the way of this flow, the cold flows unhindered across western Europe, affecting England in the North and the Sahara in the South. This cold flow has become the source of the late spring blizzards. Its been called "The Beast from the East."

The agricultures of Russia and Western Europe are severely threatened by the Beast, probably as severely as the agricultures of Canada and the Northern USA are threatened by the 'Beast' from the North.

Only the agricultures of China, Indonesia, and India are somewhat shielded from the large, cold northern air mass, by high mountain ranges standing in the way.

Part 7 - Exit from the Boundary Zone - Plan-B

Part 7

Exit From the Boundary Zone

Plan-B

Part 7 - Exit From the Boundary Zone - Plan-B

The world food crisis becomes unimaginable

Canada 36 million people
corn 14 - wheat 30

The USA 326 million people
corn 370 - wheat 56

European Union 508 million people
corn 61 - wheat 157

Russia 146 million people
corn 13 - wheat 59

China 1.4 billion people
corn 215 - wheat 126 - rice 206

India 1.3 billion people
corn 27 - wheat 96 - rice 157

Vulnerable agriculture regions

corn 458 - wheat 302 (in millions of metric tons)

1.016 billion people

Worldwide Major Grains Production

Corn 1.036 billion tons
Wheat 759 million tons
Rice 487 million tons

When agricultures fail, the world food crisis becomes unimaginable.

Take wheat and corn production, as an example, which is the main food crop, except for rice in the rice producing regions. What happens when the 4 major production centers that are located in the vulnerable agriculture regions, experience catastrophic crop failures as the result of the now increasing climate collapse? Over a billion people become affected thereby. Will they simply migrate away into warmer areas where agriculture is less vulnerable to the effects of the cooling Earth, such as in China and India, or Africa?

A mass migration on this scale is absolutely unimaginable, even into Africa; nor is it physically possible. Neither is it physically possible for the remaining large agricultural centers, China and India, to spare enough food to nourish the collapsing Western World. China and India will most likely suffer major crop losses too, in the collapsing global climate.

So what is the answer? The answer is Plan-B. No other option exists. The agricultural collapse in the climate-vulnerable areas cannot be avoided. The collapse is already beginning, even while we are only half-way into Phase-1 of the solar weakening in the boundary zone, with 15 more years to go till Phase-2 kicks in, when things get worse.

Plan-B doesn't change the climate

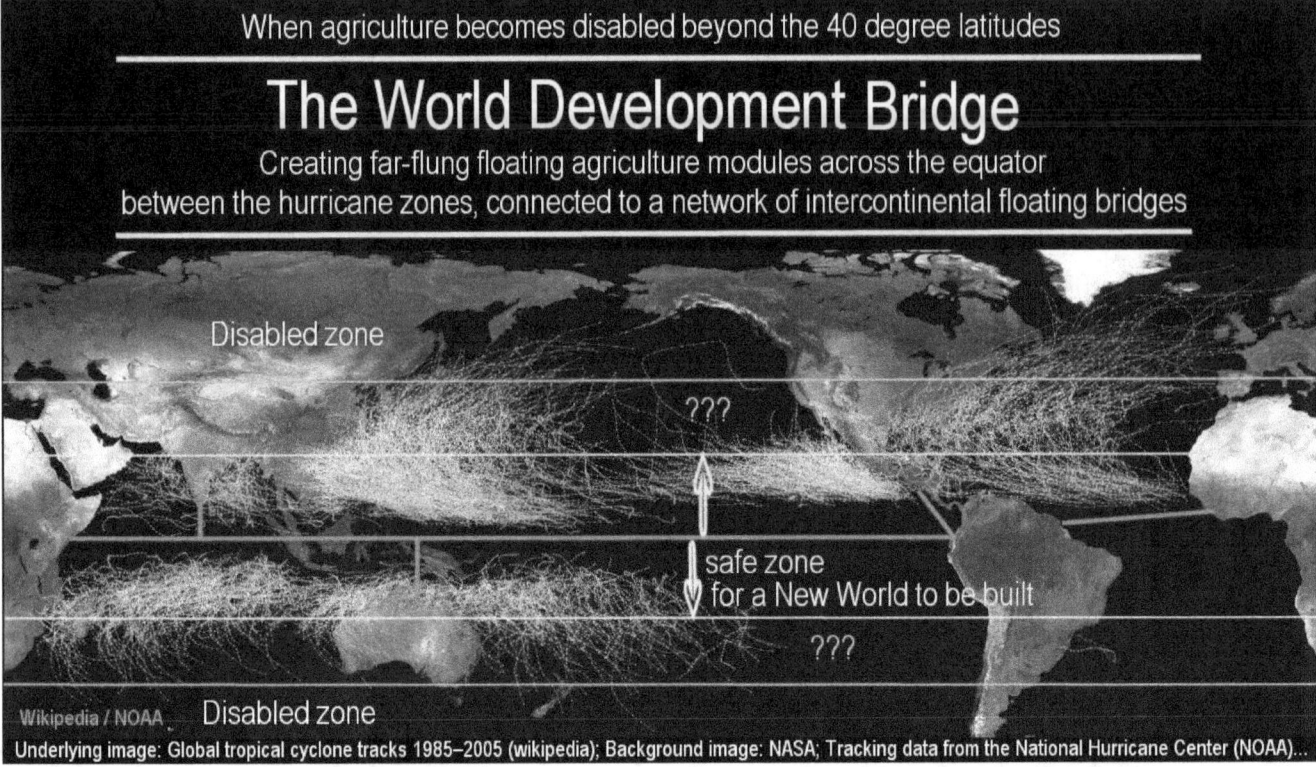

Plan-B doesn't change the climate, but it enables us to live with it, and flourish in-spite of it. Unfortunately, Plan-B requires a platform that we find hard to build. It requires the cooperative participation of all nations, to build the technological infrastructures. That's where we find our greatest challenge. We are not good in the department of cooperative participation. The West champions the opposite, imperial stealing. We fight wars not to cooperate, but to subdue, destroy, and plunder.

Let's consider the Canadian Wheat Board

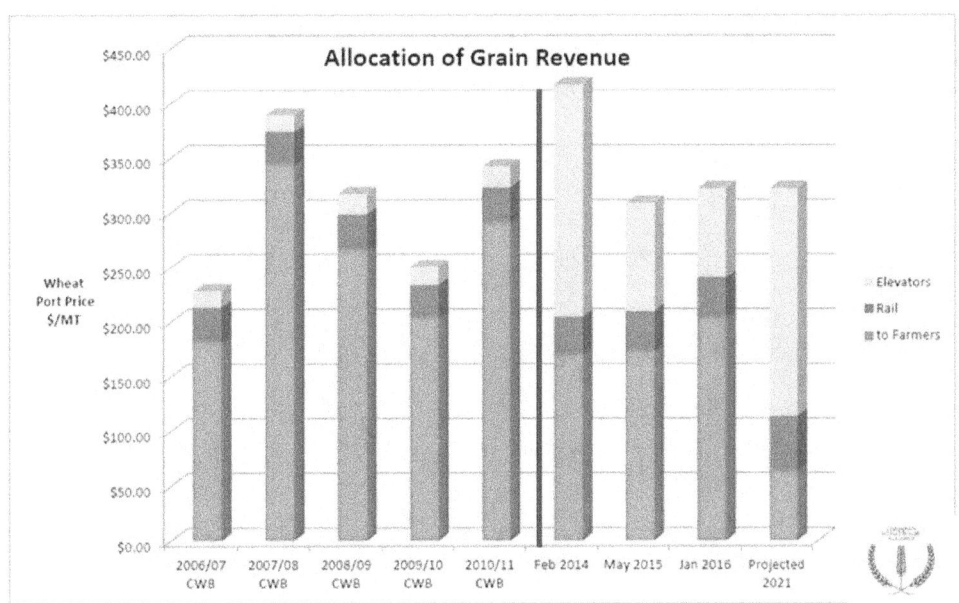

Sources: CWB Audited Statements, CWB Market Newsletter Feb 2014, Canadian Transportation Agency MRE determinations, Canadian Grain Commission posted elevator tariff rates, Agriculture and AgriFood Canada Weekly port price summaries for # 1 CWRS wheat Vancouver.

2021 projection based on University of Sask. Ag Economist J. Nolan —assumes consolidation of grain handlers - also MRE removed.
http://www.cwbafacts.ca/2016/02/for-your-viewing-pleasure/

http://www.cwbafacts.ca

Let's consider the Canadian Wheat Board as a tiny example. It operated as a government-run institution in support of the farmers. It assured that the bulk of the sales revenue was returned to the farmers, the producer of the product. Its purpose was to assure a healthy farming industry.

131

The government caved in and surrendered

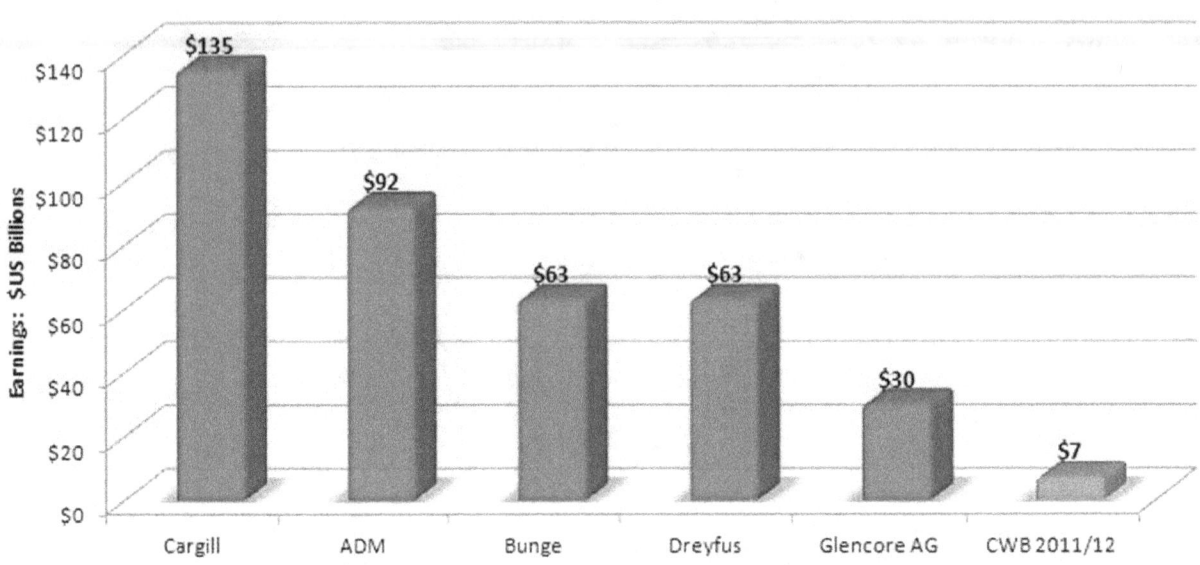

Evidently under relentless pressure, the government caved in and surrendered the Wheat Board into the hands of some of the large international marketing cartels.

When a nation no longer protects its farm industry

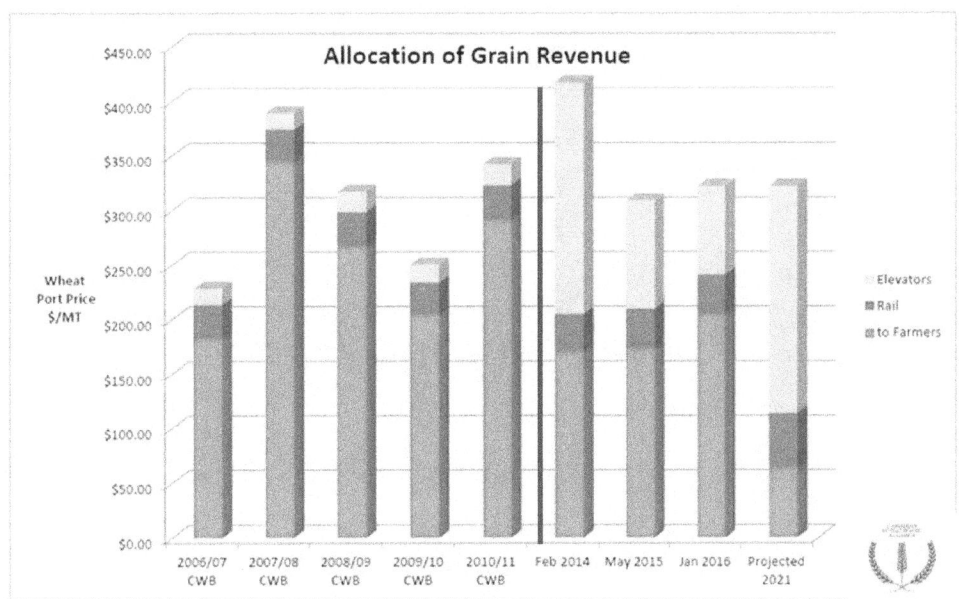

Sources: CWB Audited Statements, CWB Market Newsletter Feb 2014, Canadian Transportation Agency MRE determinations, Canadian Grain Commission posted elevator tariff rates, Agriculture and AgriFood Canada Weekly port price summaries for # 1 CWRS wheat Vancouver.

2021 projection based on University of Sask. Ag Economist J. Nolan —assumes consolidation of grain handlers - also MRE removed.
http://www.cwbafacts.ca/2016/02/for-your-viewing-pleasure/

http://www.cwbafacts.ca

As the result, the share for the farmers, of the sale price, was dramatically reduced from 2014 on, by private profit demands. The tragedy is, that when a nation no longer protects its farm industry, it gambles with the foundation for its existence. This letting go of what is vital for the sake of profit in the shadow of corruption is the kind of disease that stands in the way of universal cooperative engagement. In other words, the general welfare principle is dead, while this principle is precisely what makes a Plan-B feasible.

Plan-B anyone?

Plan-B anyone?

Plan-B anyone?

The climate collapse is already beginning to bite

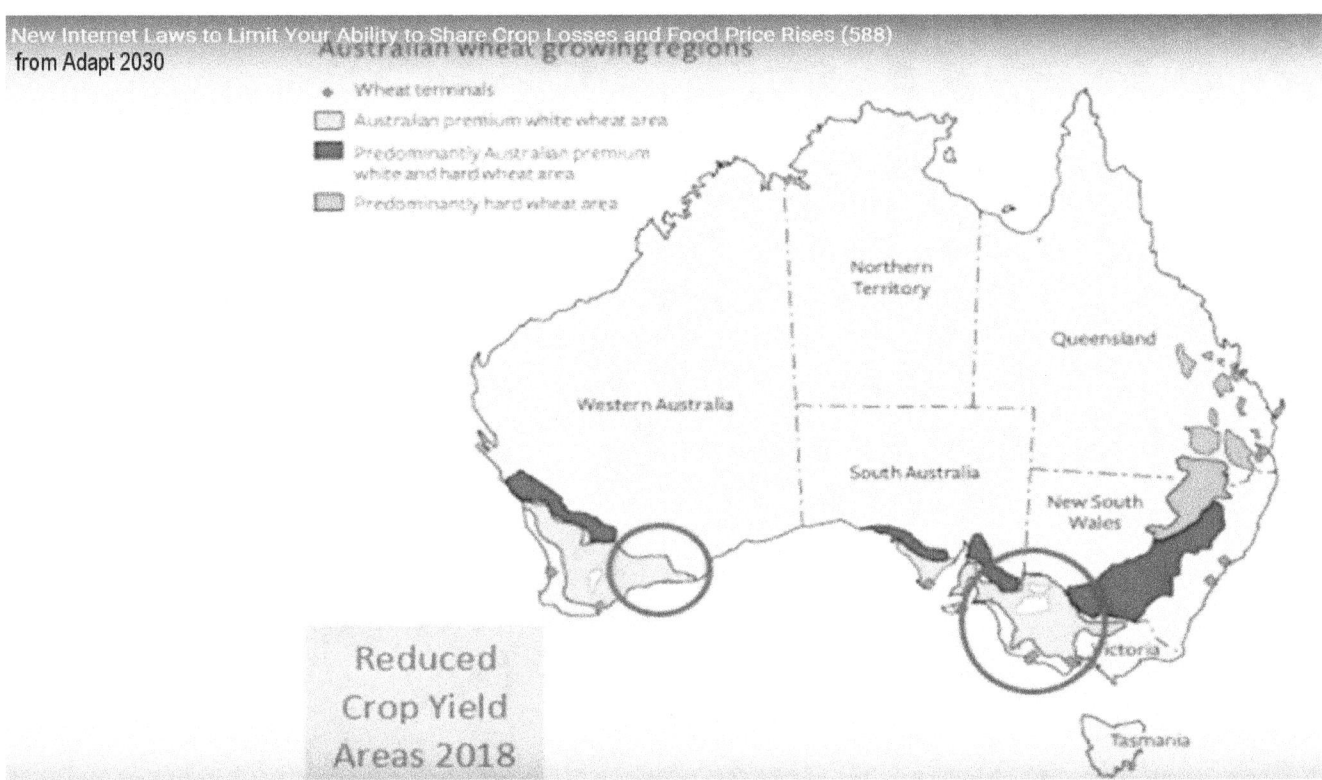

The climate collapse by the weakening Sun is already beginning to bite, and in some places quite hard. Australia lost 30% of the national wheat crop in 2017, due to drought and excess heat, which are some of the fringe effects of increasing cosmic-ray flux that affects cloud forming, cloud rainout, and reduced water vapor in the air that makes up 97% of the moderating greenhouse effect.

In the same year Europe suffered a major loss of vegetable crops due to adverse climate issues, which had led to vegetable rationing in some areas.

The complete collapse of agriculture

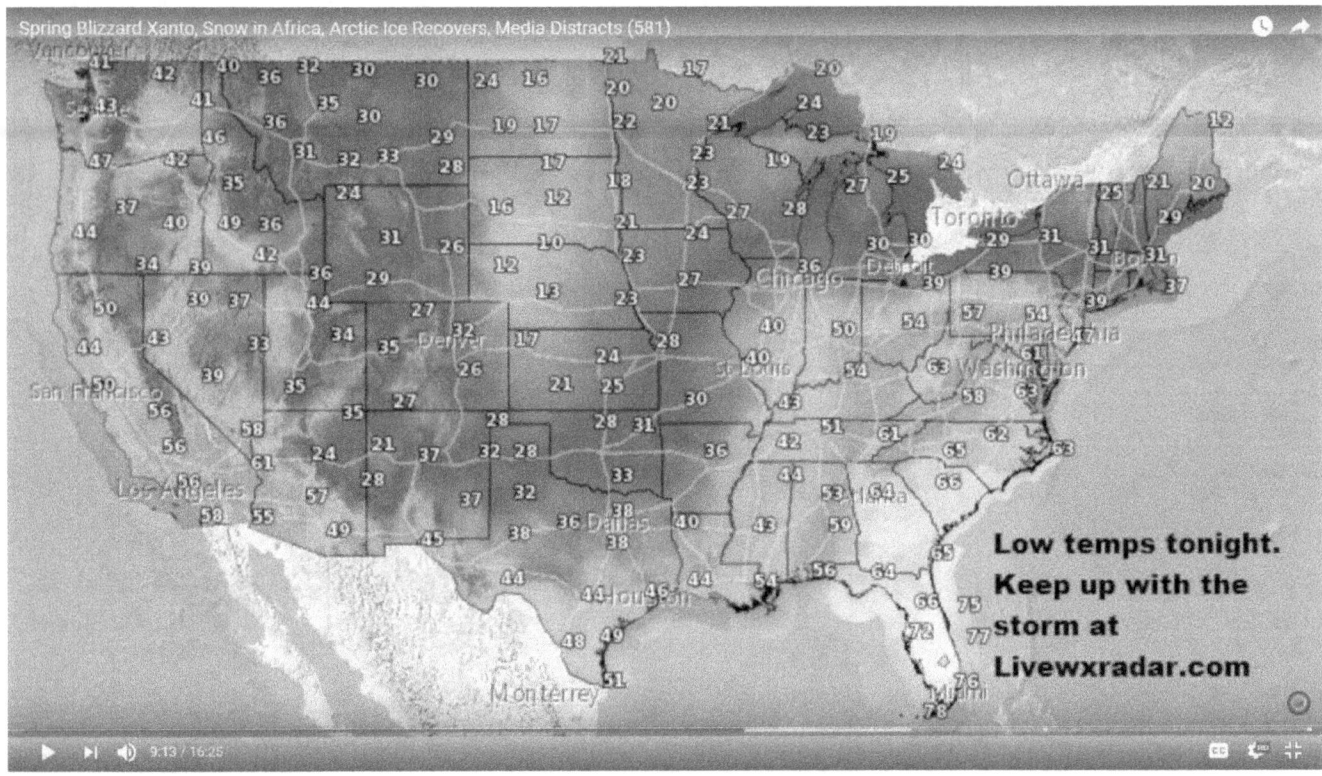

On the American continent, the late-spring blizzard "Xanto" in April 2018, across the American grain belt, will most likely also result in significant crop losses in volume and quality.

With these large effects already happening, and the Earth getting colder year after year for several more decades to come, the complete collapse of agriculture in the climate vulnerable regions will happen unavoidably within the next 15 years.

Only the effects of the collapse can be avoided

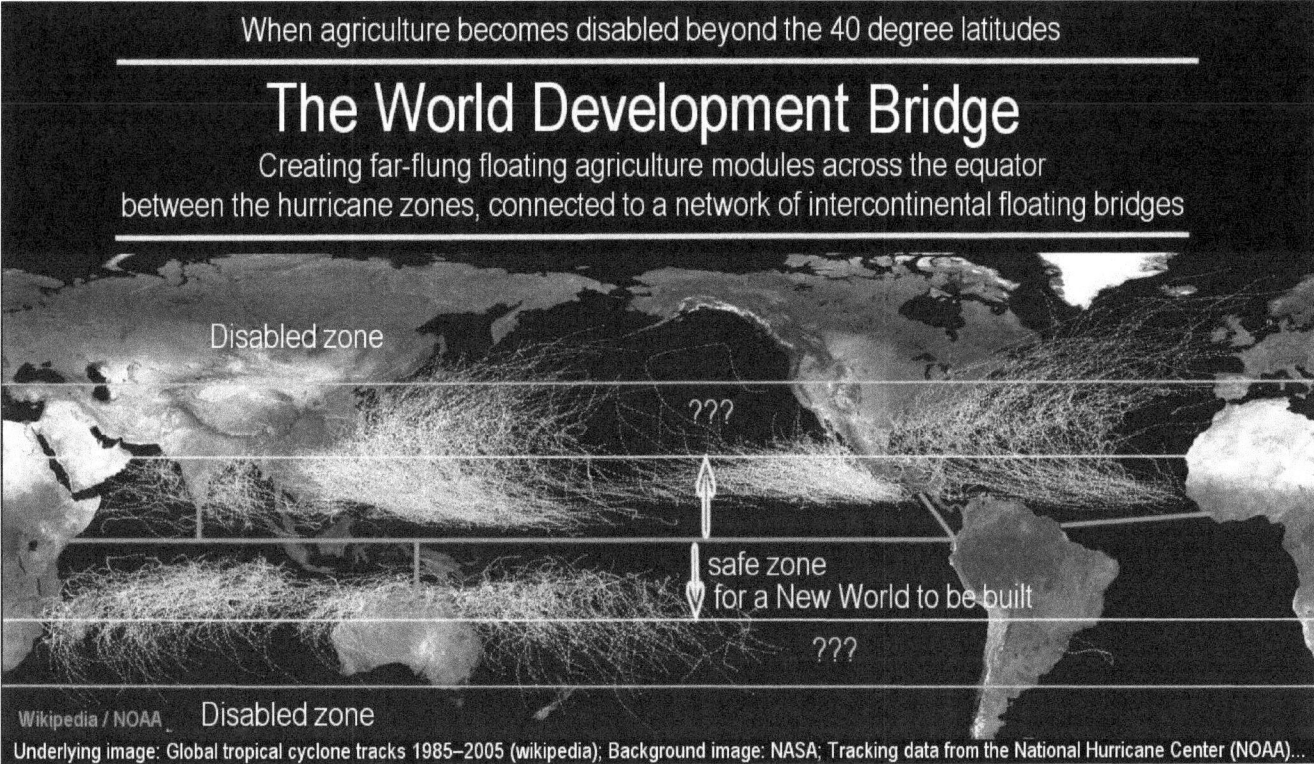

Only the effects of the collapse can be avoided by implementing a Plan-B option, such as the one that creates vast new technologically protected agricultures afloat on the equatorial sea. Such a Plan-B option is quite easily implemented with large-scale automated industrial production methods and the large scale use of basalt as feed-stock, and high-temperature nuclear energy to power the process.

However, a new type of thinking is required in the world for this large-scale critical option to become acceptable. Society has stopped thinking too long ago in terms of its renaissance and industrial revolution potential. Its thinking has become too small and encumbered with imagined limits and boundaries to what is achievable.

Becoming Unbound from Small Thinking

Becoming Unbound
from
Small Thinking

Becoming Unbound from Small Thinking

The floating World Bridge

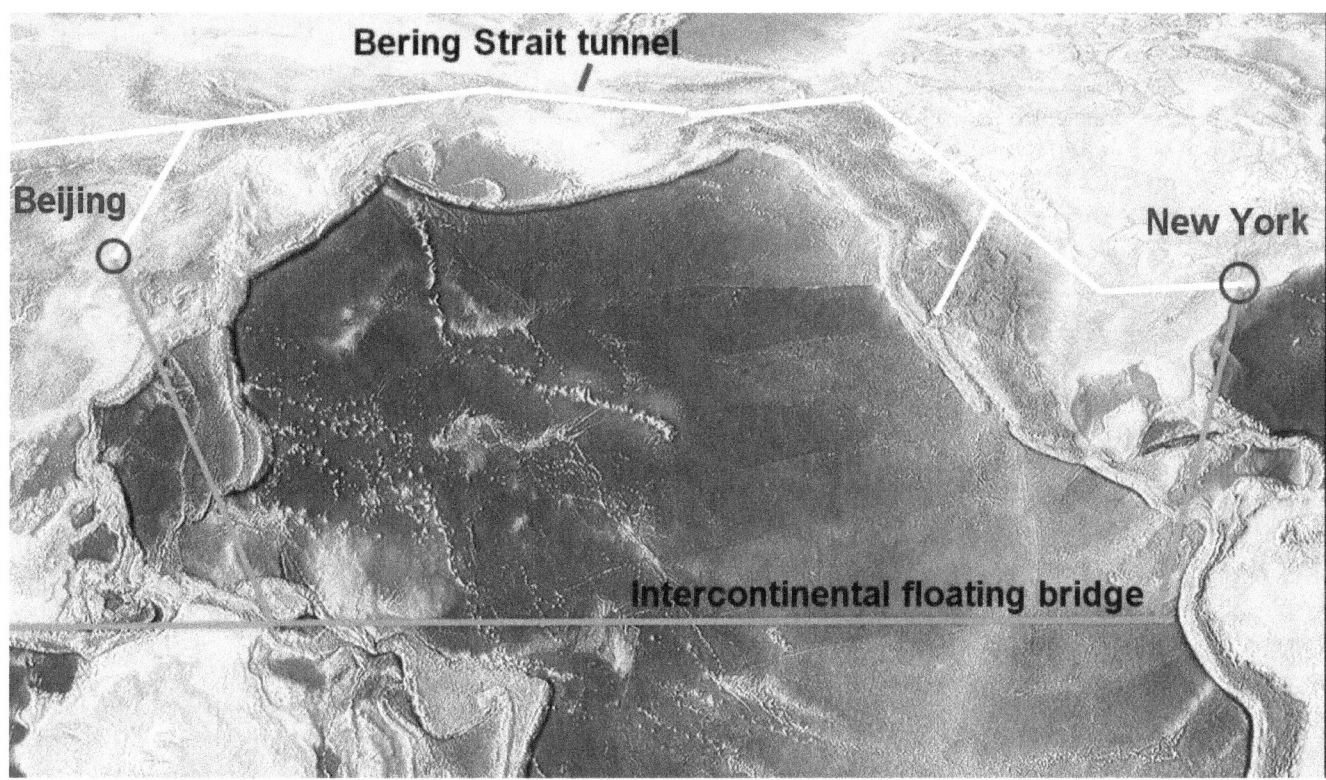

The floating World Bridge as a concept, that links the continents across the tropical seas, appears totally impossible to accept in today's thinking, while it is actually more efficient than any other option that is being contemplated, such as the 50 miles long Bearing Strait tunnel link between the Eurasian and American continents.

The Bering Strait under-sea-tunnel link

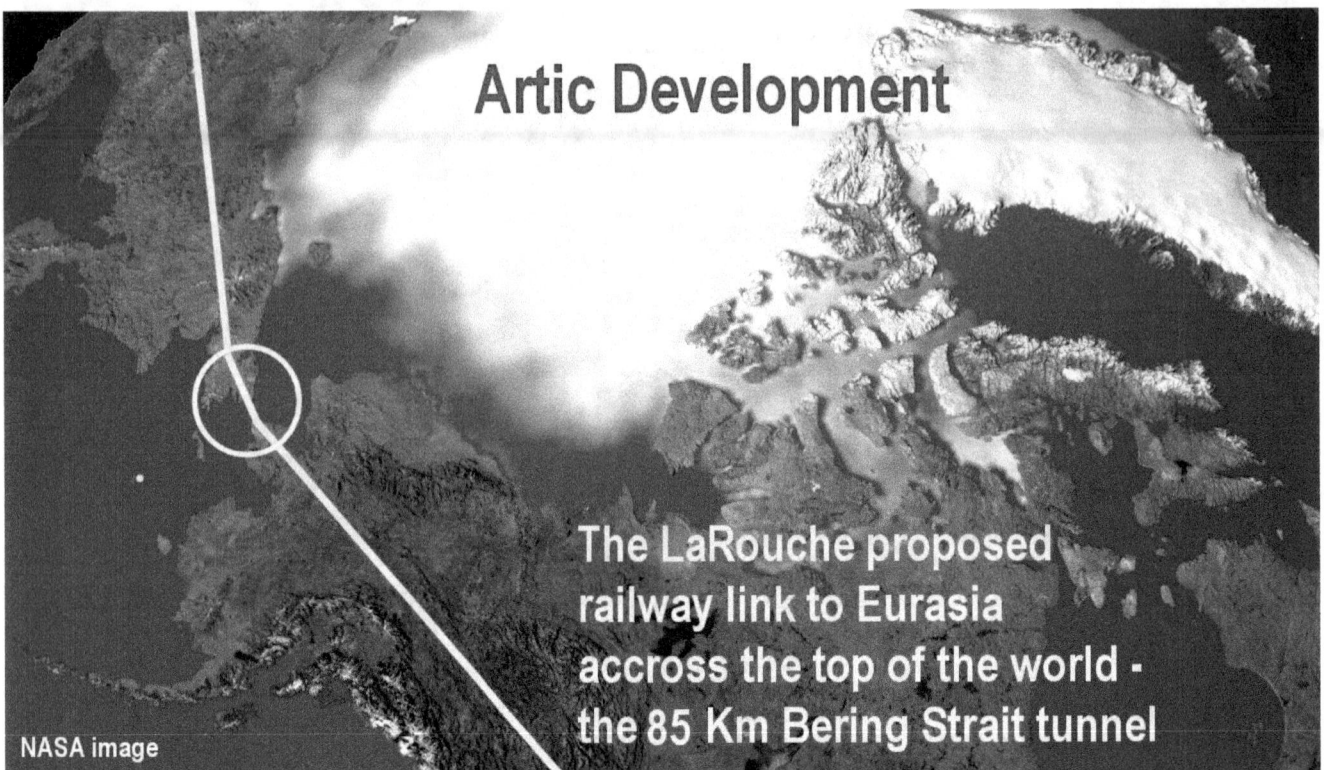

The Bering Strait under-sea-tunnel link has been promoted for years. It would take a dozen years to build the 50 miles long tunnels. And it would require thousands of kilometers of new railway to be built, to link up with it, with much of its across difficult terrain with harsh climate conditions. An enormous amount of manual labor would be required to build such a project.

The floating intercontinental World Bridge

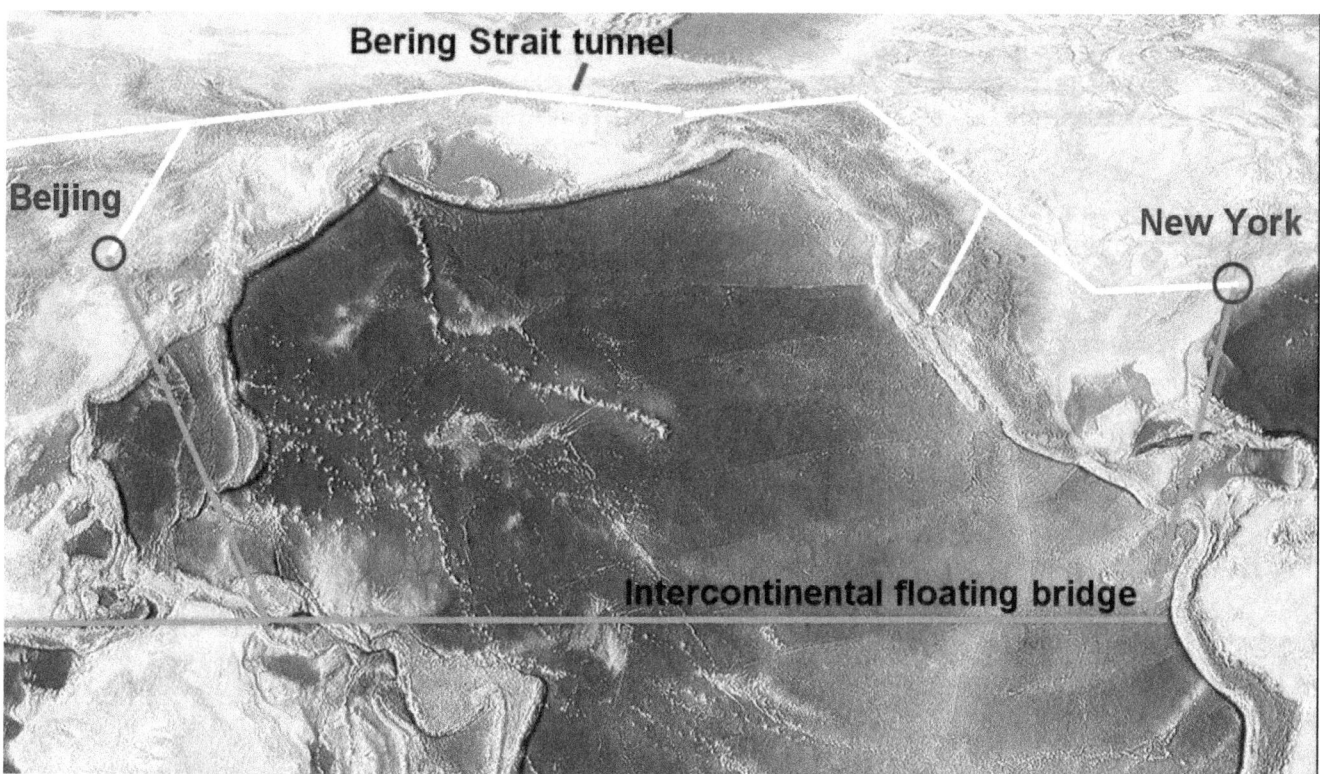

The floating intercontinental World Bridge, in comparison, can be easily built with uniform modules produced in automated industrial production, placed with automated equipment without the need for large labor brigades. It may well be that the 12,000 Km long world bridge across the Pacific will take less time to construct than it would take to construct the 50 miles undersea tunnel across the Bering Strait. The speed of the construction for the World Bridge would then be determined by the scale of the constructing industry. In this case, it is just a matter of scale. The principle of automated construction can be applied to any scale.

The problem that's holding us back

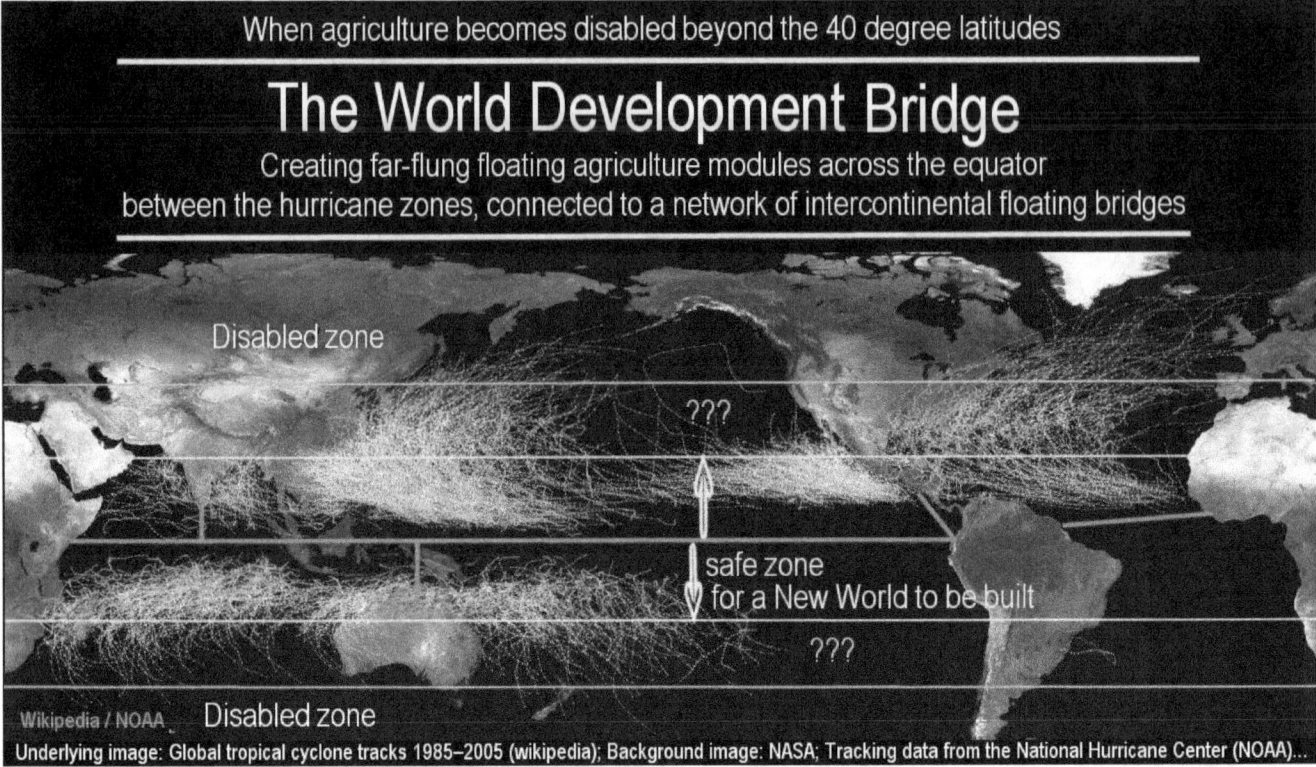

This means that the problem that's holding us back so severely that nothing has been started yet, is nothing more than a bad case of small-scale thinking.

Unbounded Productivity: The Automation Principle

Unbounded Productivity

The Automation Principle

Unbounded Productivity: The Automation Principle

A simple example, in principle

Let me present you a simple example, in principle, what can be accomplished when automated industrial production is being applied.The example shown here is that of the automated track laying process for the new high-speed railway in Northern China, a 140 km rail link between the city Datong and the city Zhangjiakou that will host the 2022 Winter Olympics.

The machine lays down the concrete ties and 500 meter rails

The machine lays down the concrete ties and positions and fastens the 500 meter rails with the needed precision for trains travelling 250 Km/hour.

In the past, large labor brigades had been needed

In the past, large labor brigades had been needed to carry everything into place. Now, very few people are found on the construction site, and those merely supervise the operation of the machine.

The machine lays down 2,000 meters of rail per day

Even the supplies are automatically delivered from the transport cars to the placing machine, by a carrier running overhead on side rails, as the machine keeps on going and going and lays down 2,000 meters of rail per day.

If one applies the automation principle to the World Bridge construction

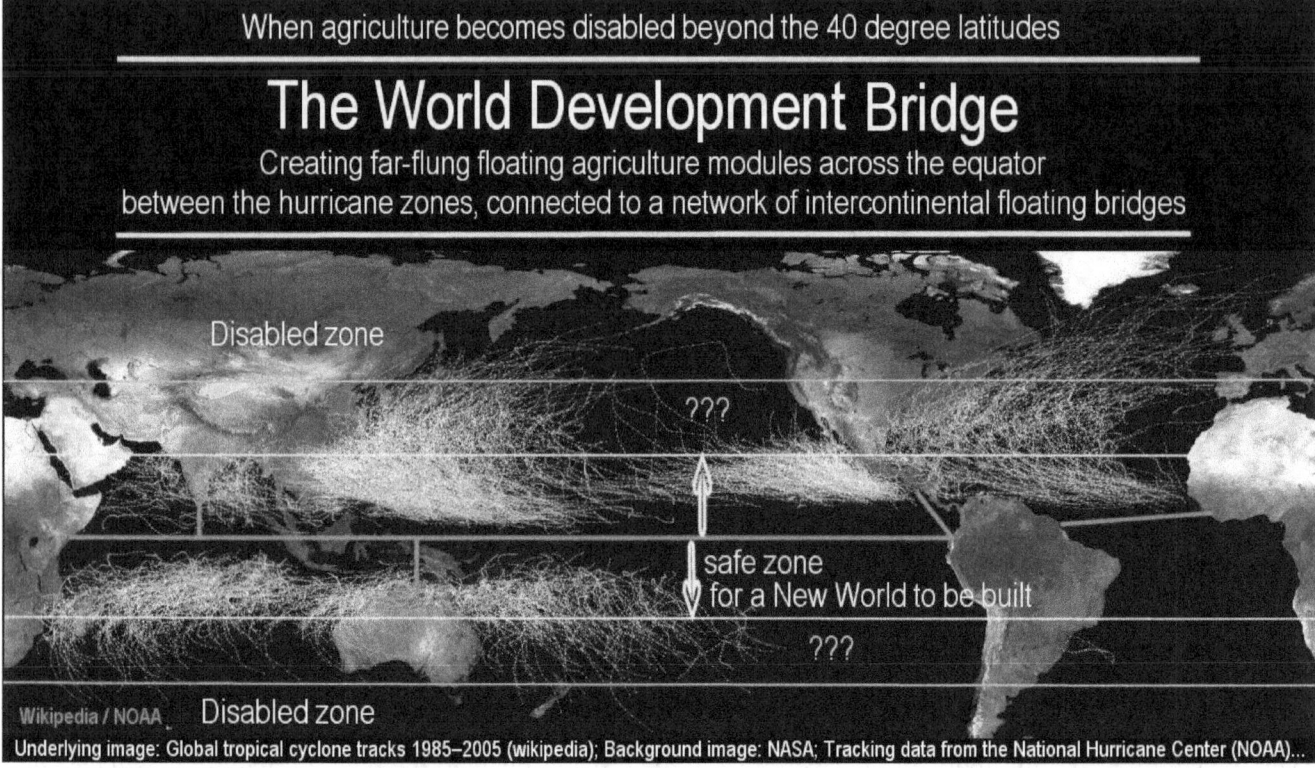

If one applies the automation principle to the larger scale of the World Bridge construction, with agricultural modules extending from them, including the needed cities for them, the entire 12,000 Km world bridge project spanning the sea between the continents, could be similarly be completed, without anyone putting a spate into the ground so to speak, and could be completed in just a few years if the construction was carried out in numerous sections simultaneously with the producing industries for the modules being afloat themselves..

Also, as the world-bridge was being built, the agricultural modules linked to it would bring large scale food-production online, almost from the start, to compensate for the climate-related agricultural losses in the Northern regions. By this Plan-B approach the world wouldn't run low on food when the agricultures in the cold northern regions begin to collapse, which may happen in 5 to 15 years..

The world-bridge agriculture, as the world bridge itself and its cities, would be cooperative built by humanity at large, and be never owned by anyone. It would be a joined development product of the entire world for itself, as no nation is ultimately not affected by the ongoing climate collapse, especially in later years when the Ice Age begins that the boundary zone takes us into.

As the new agricultures develop, thousand of new cities will be linked to them for the people that operate in the infrastructures and processes, and the industries that support them. And of course, living in the new world would be rent free. The concept of property 'lords' enslaving humanity, would simply fall away. The focus would then be on developing the human potential as the greatest treasure that a society has. A type of money-free platform will likely emerge in due course where advancing culture, human progress, and productive technologies go hand in hand.

The New World: An Unbounded World

The New World

An Unbounded World

The New World: An Unbounded World

This is the type of Plan-B option that comes into view

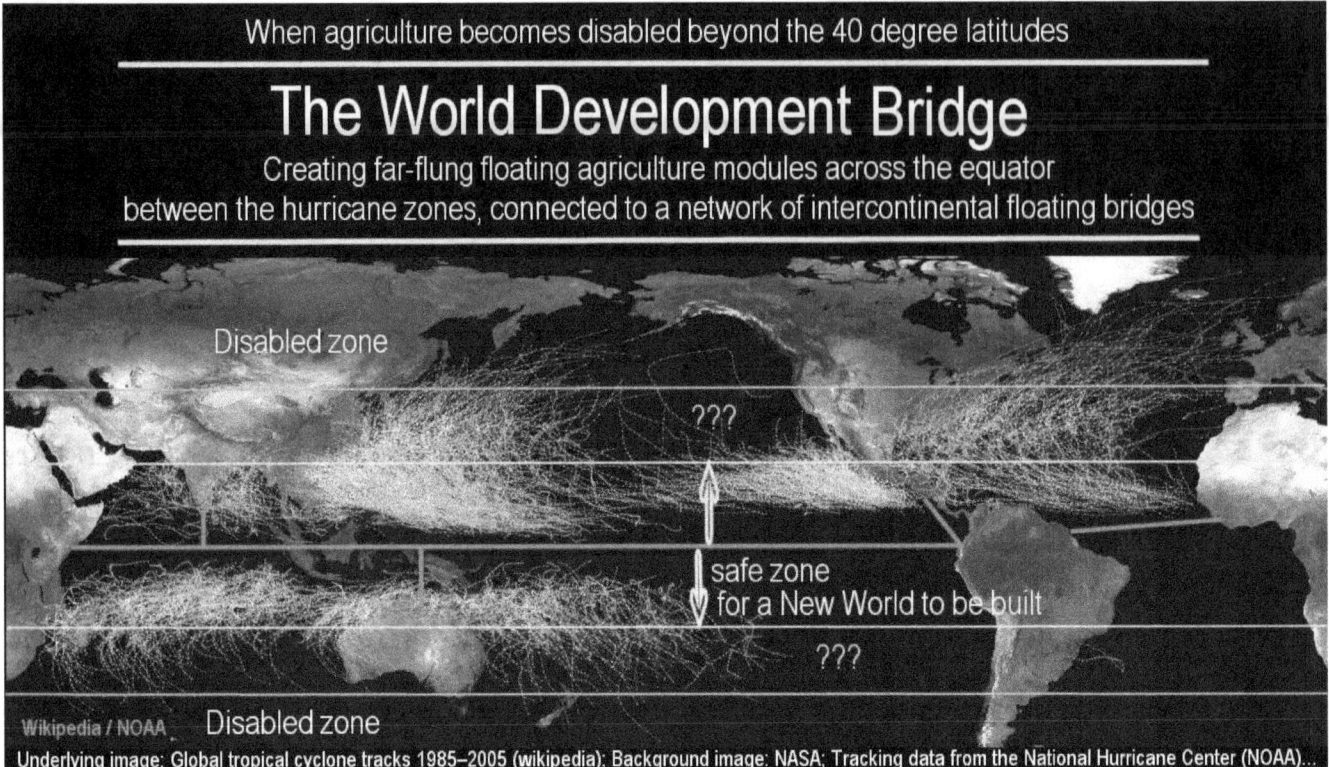

This is the type of Plan-B option that comes into view when the boundaries of small-minded thinking fall away. The Plan-B world will then become created as an unbounded world.

Actually, the unbounded part isn't optional. Any Plan-B option of the required scale and with the necessary speed of implementation, can only be created on an unbounded basis, unbound from money and property concepts, and society's slavery to them. A new age will then begin, designed from the start as an unbounded age of industrial and cultural revolutions, because without these, no form of Plan-B is actually possible, even to contemplate.

Should a Plan-B not materialize, the Plan-A option - to do nothing - would prevail, and its song "let the people die" becomes the requiem for the self-termination of humanity when the Ice Age phase shift happens in the 2050s.

The Ice Age will resume by the principles of the cosmic dynamics

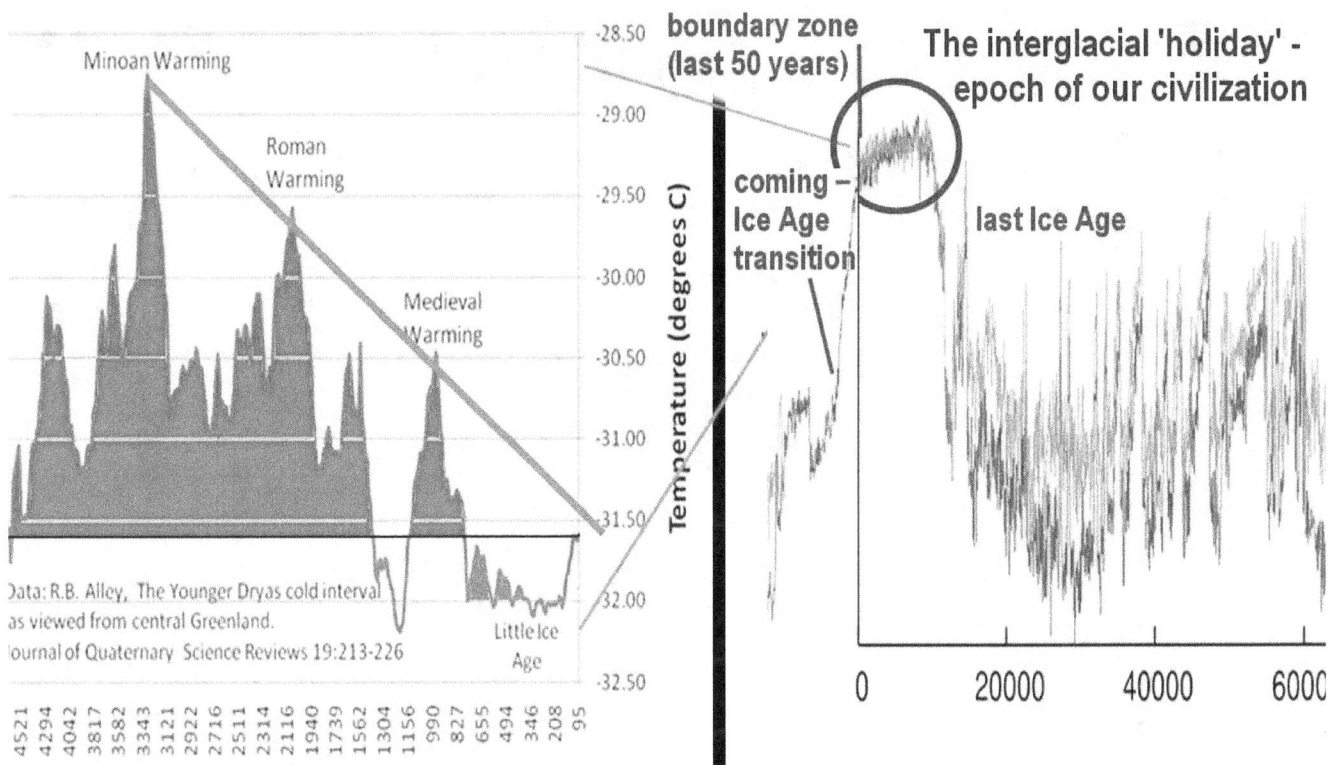

The Ice Age will resume by the principles of the cosmic dynamics. These, we have explored and measured, extensively, with satellites in space, measurements on the ground, in the air, in ice cores, and with ground based radio telescopes. The ending of the current interglacial epoch has been well explored and it is evidently extensively understood. There is little chance that the Ice Age phase shift will not happen in the way it always has.

The only uncertainty that remains in the entire Ice Age arena

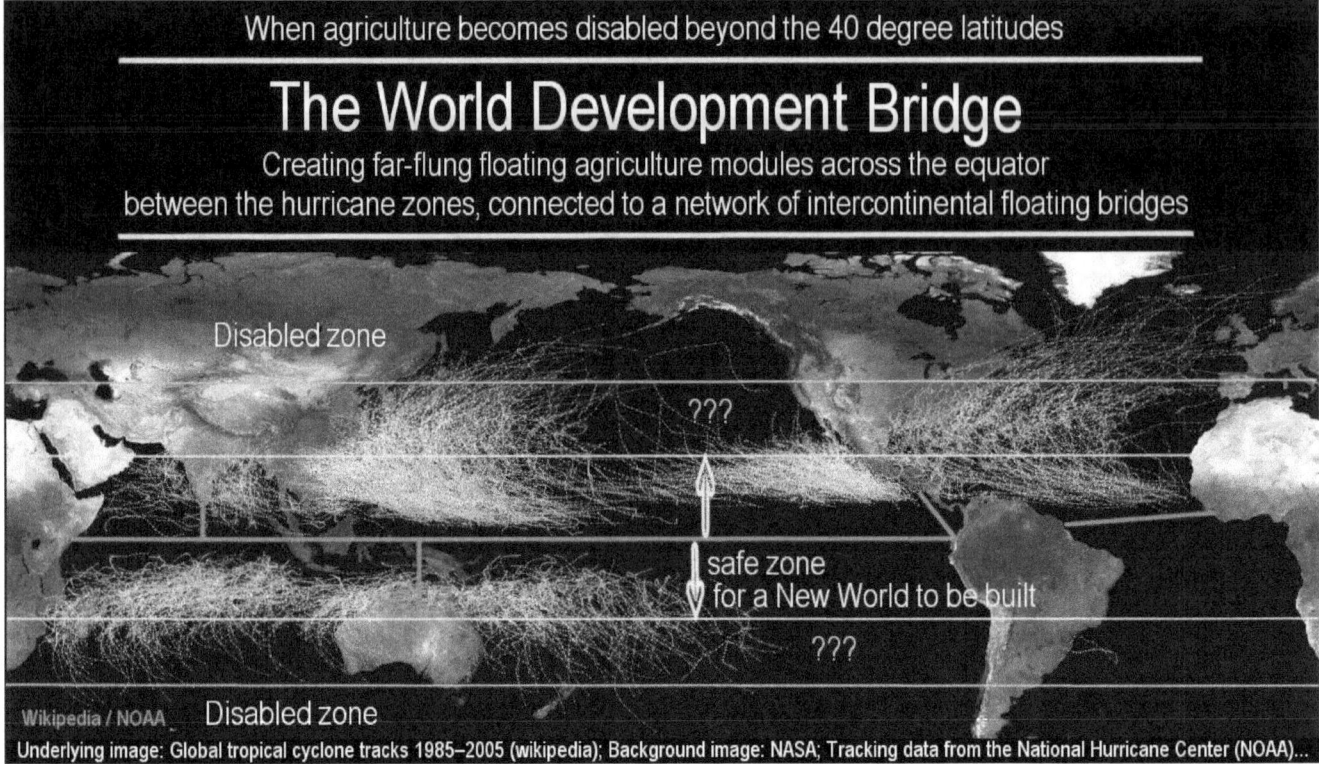

The only uncertainty that remains in the entire Ice Age arena, is how humanity will respond to what has been measured and is known; whether it will respond with Plan-A or Plan-B. That's only uncertainty remaining at this time.

At the present, Plan-A is the ruling king

At the present, Plan-A is the ruling king.

Part 8 -
Deadly Blocking Factor: The Global Warming Hoax

Part 8

Deadly Blocking Factor

The Global Warming Hoax

Part 8 - Deadly Blocking Factor: The Global Warming Hoax

Whether humanity will choose Plan-B and survives

PARIS2015
UN CLIMATE CHANGE CONFERENCE
COP21·CMP11

Poster of the Climate Conference.
Licensed under Fair use via Wikipedia

COP 21: Heads of delegations by GUSTAVO-CAMACHO-GONZALEZ - Licensed under CC BY 2.0 via Commons by Presidencia de la República Mexicana -delegates

Whether humanity will choose Plan-B and survives, depends on its success in clearing its most deadly blocking factor out of the way. This factor isn't nuclear war, or financial collapse, or terrorism, narcotics, migration, or even trade disputes.

The blocking factor is the doctrine of manmade global warming

The blocking factor is the doctrine of manmade global warming that every nation in the world inherently bows to. It was designed from the very beginning to stand as a blocking barrier against any form of Plan-B global development in response to the coming Ice Age.

The Global Warming Doctrine was invented in 1975 to overturn the concerns in the scientific community of how the world needs to prepare itself for the near Ice Age consequences.

The 'writing' was 'on the wall'

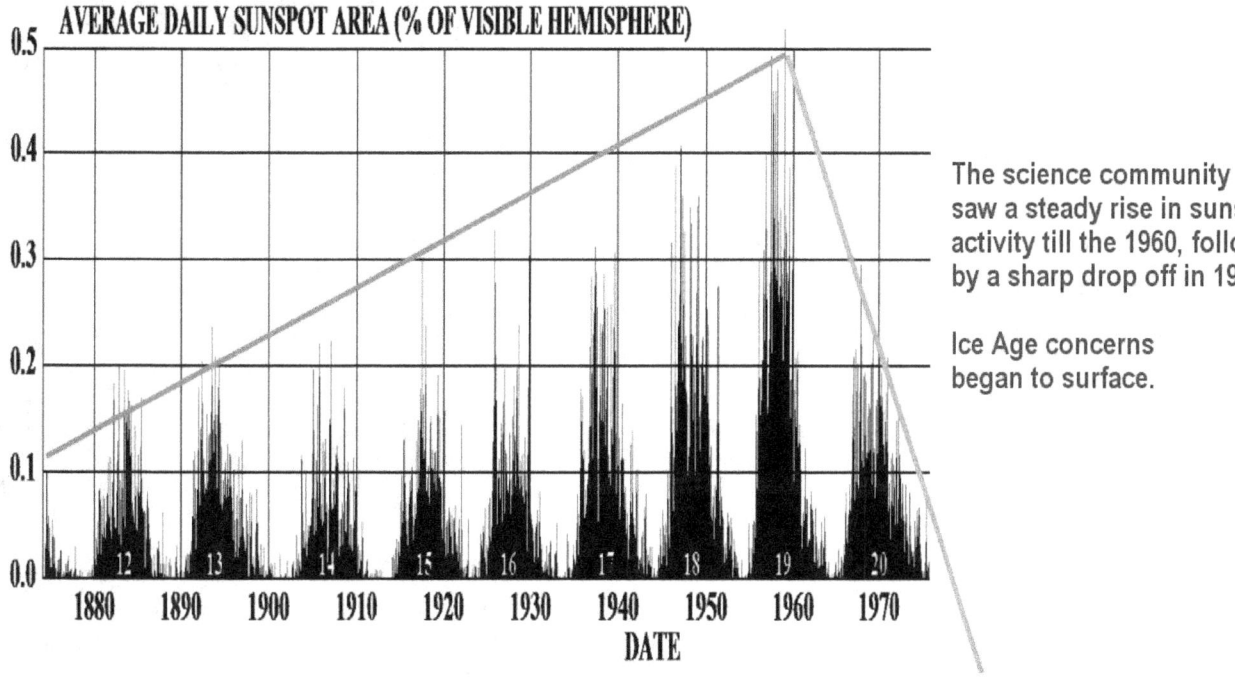

AVERAGE DAILY SUNSPOT AREA (% OF VISIBLE HEMISPHERE)

The science community saw a steady rise in sunspot activity till the 1960, followed by a sharp drop off in 1970.

Ice Age concerns began to surface.

http://solarscience.msfc.nasa.gov/

The science community saw a steady rise in sunspot activity till the 1960s, and saw it followed by a sharp drop off in the 1970s. Ice Age concerns began to surface. But how to prepare for it?

Any preparation for an Ice Age on the worldwide scale, would required massive economic development. The 'writing' was 'on the wall', so to speak. But the timing was bad.

The ruling elite of the imperial oligarchies was aiming for depopulation at the time. They feared economic development like the plague. It was to be prevented at all cost to preserve the ruling imperial system. For a few pennies they hired a group of scientists who thought up a lie, quick and slick, and it was potent.

The preposterous claim that carbon dioxide will broil the Earth

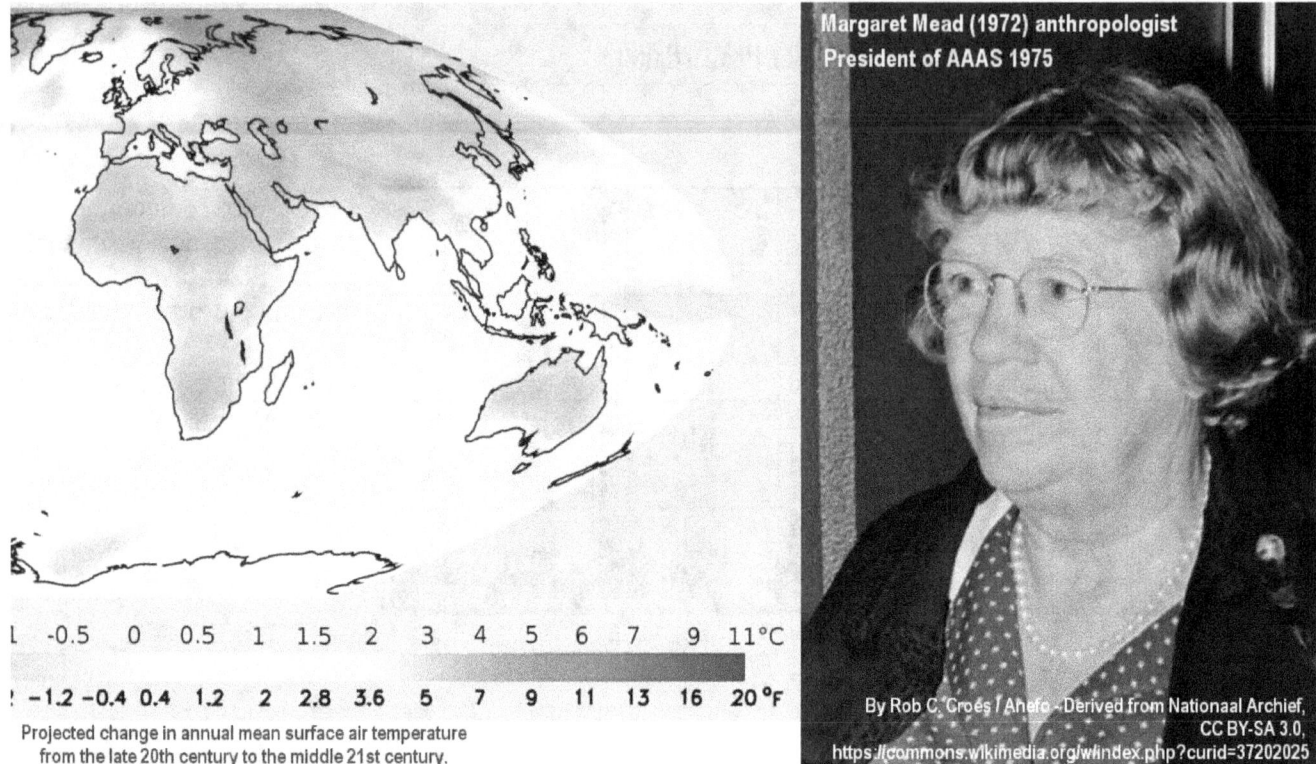

| L | -0.5 | 0 | 0.5 | 1 | 1.5 | 2 | 3 | 4 | 5 | 6 | 7 | 9 | 11°C |
| : | -1.2 | -0.4 | 0.4 | 1.2 | 2 | 2.8 | 3.6 | 5 | 7 | 9 | 11 | 13 | 16 | 20 °F |

Projected change in annual mean surface air temperature
from the late 20th century to the middle 21st century.

Margaret Mead (1972) anthropologist
President of AAAS 1975

By Rob C. Croés / Anefo - Derived from Nationaal Archief,
CC BY-SA 3.0,
https://commons.wikimedia.org/w/index.php?curid=37202025

"The preposterous claim that human-produced carbon dioxide (by our living on earth) will broil the Earth, melt the ice caps, and destroy human life, was dished up in 1975 at a conference in Research Triangle Park, North Carolina, organized in 1974 by the influential anthropologist Margaret Mead, president of the American Association for the Advancement of Science (AAAS). *1 (continued)

Note: The Research Triangle, commonly referred to as simply The Triangle, is a region in the Piedmont of North Carolina in the United States, anchored by three major research universities North Carolina State University, Duke University, University of North Carolina at Chapel Hill, (Wikipedia)

The strategy was to sow enough fear

"...The strategy was to sow enough fear of man-caused climate change to force global cutbacks in industrial activity and halt Third World development. Mead's leading recruits at the 1975 conference were climate-scare artist Stephen Schneider, population-fear biologist George Woodwell, and the current AAAS president John Holdren—all three of them disciples of Malthusian ideologist Paul Ehrlich, author of The Population Bomb. The discussion focussed on the absurd choice of either feeding humanity or 'saving the environment.'*1

Note*1 Based on: 1975 Endangered Atmosphere' Conference: Where the Global Warming Hoax Was Born

by Marjorie Mazel Hecht, in the June 8, 2007 issue of Executive Intelligence Review.

https://larouchepub.com/other/2007/sci_techs/3423init_warming_hoax.html

The right balance between being effective and being honest

"Global Warming proponent Stephen Schneider"

One of the original promoters of the global warming doctrine

Bill Rose/Michigan Technological University

Global warming proponent Stephen Schneider: ". . .[W]e have to offer up scary scenarios, make simplified, dramatic statements, and make little mention of any doubts we may have. Each of us has to decide what is the right balance between being effective and being honest."

November Edition, 1997, of magazine: 21st Century Science and Technology

"We have to offer up scary scenarios, make simplified, dramatic statements, and make little mention of any doubts we may have.

"Each of us has to decide what is the right balance between being effective and being honest."

"We have to offer up scary scenarios, make simplified, dramatic statements, and make little mention of any doubts we may have. Each of us has to decide what is the right balance between being effective and being honest."

So spoke Global Warming proponent Stephen Schneider.

Steven Schneider, one of the original promoters of the global warming doctrine

November Edition, 1997, of magazine:

21st Century Science and Technology

Scientists were charged to back up the scares

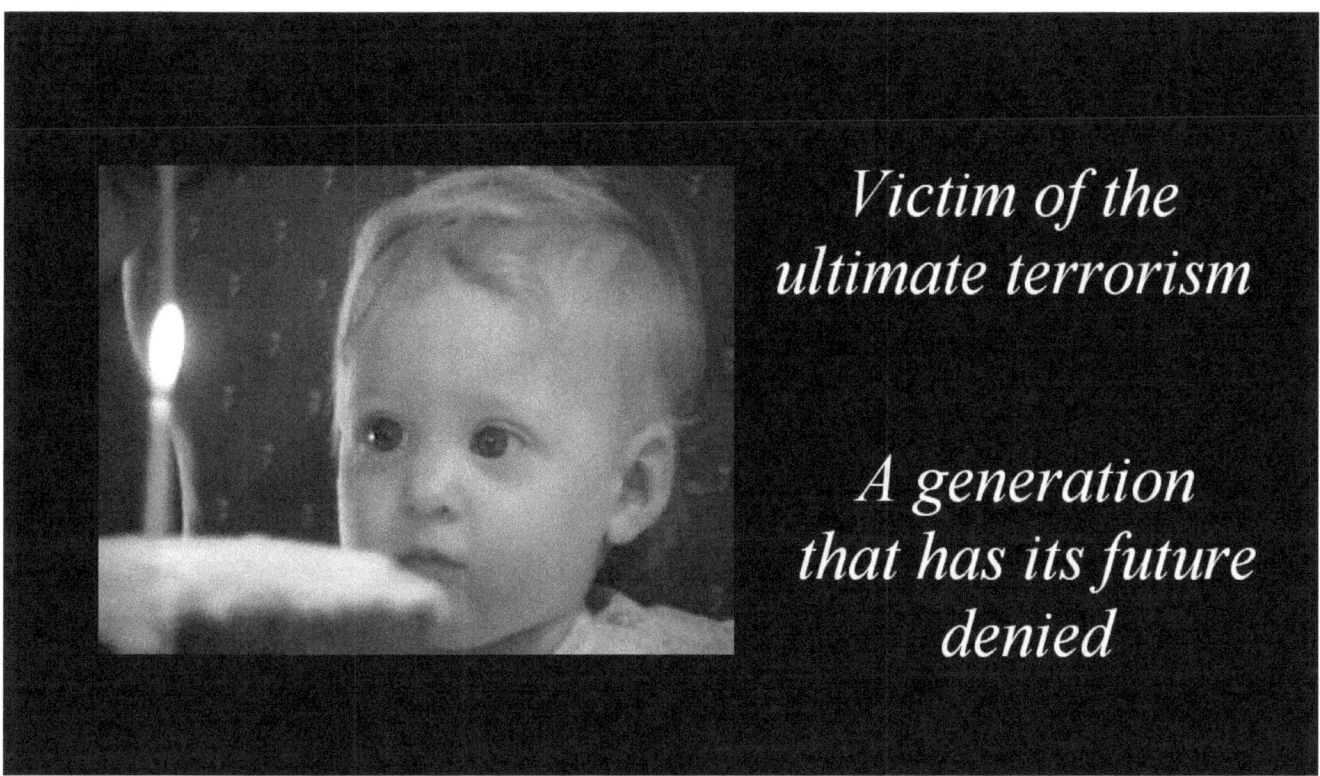

It was at the infamous government-sponsored conference in 1975 that virtually every scare scenario in today's climate hoax took root. Scientists were charged with coming up with the "science" to back up the scares, so that definitive action could be taken by policy-makers to carry out the genocide.

Global cooling—the coming of an Ice Age—had been in the headlines in the 1970s, but since it could not easily be used to sell genocide by means of getting the citizens of industrial nations to cut back on their living instead of advance their development, the Ice Age concerns were buried. *1(see previous pages)

Worldwide Science Opposition

Worldwide Science Opposition

Protesting the Global Warming false assumptions

Worldwide Science Opposition Protesting the Global Warming false assumptions

Over 50,000 signatures and statements were collected in appeals

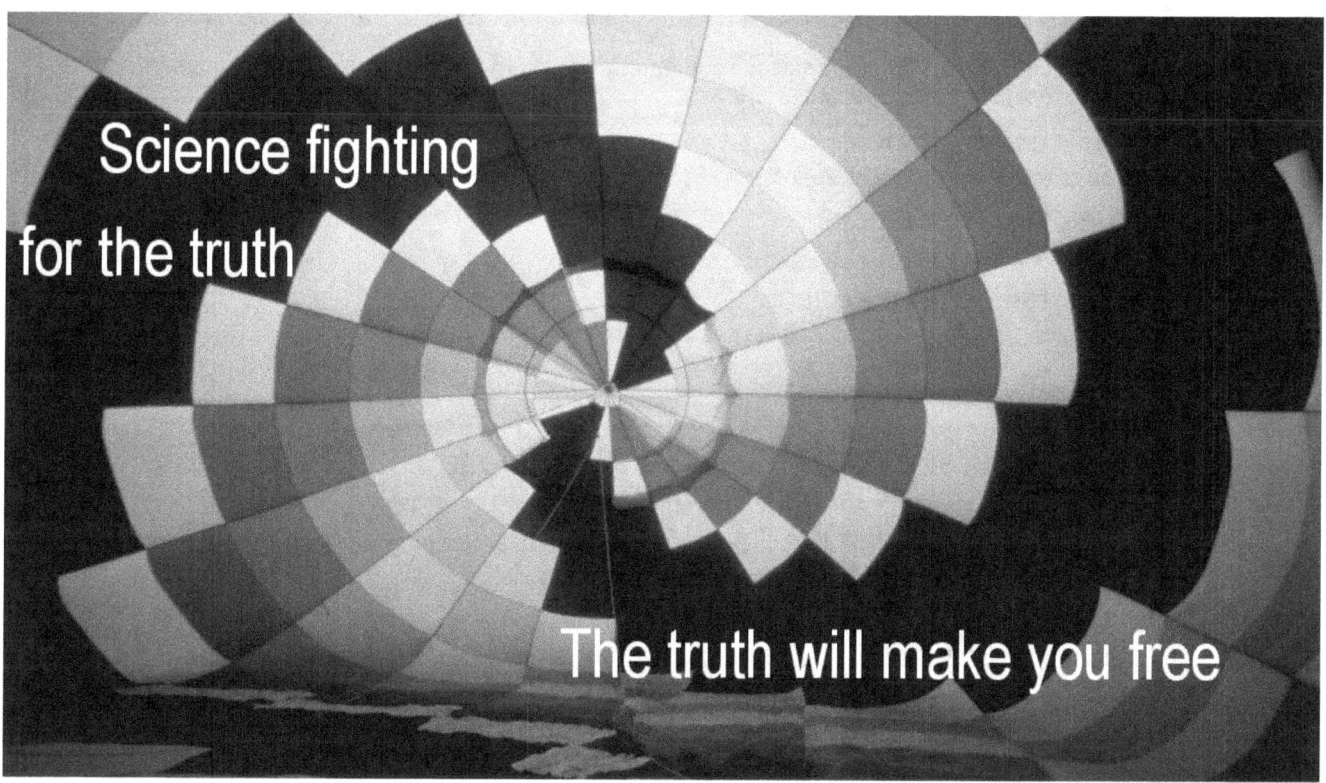

Behind the scene, the science community protested. They protested not to restore the focus on the Ice Age imperative. They merely protested against the unscientific assumptions of the manmade global warming doctrine.

The science protest movements began in 1992. Since then, over 50,000 signatures and statements were collected in appeals and petition projects from the scientific community of numerous countries, with Nobel Laureates among the respondents. But here too, the Ice Age remained out of sight and out of the discussion.

The big protests began in 1992

Over 50,000 signatures from the science community opposing the Manmade Global Warming doctrine

1992 The Heidelberg Appeal
- signed by 4000 scientists from 69 countries, including 63 Nobel Laureates

1997 The Leipzig Declaration
- signed by 110 climate specialists

1998 The Oregon Petition Project
- signed by 17,000 scientists (organized against the Kyoto Protocol)
The petition was organized and circulated by Arthur B. Robinson, president of the Oregon Institute of Science and Medicine

The Kyoto Protocol met with an 85% rejection across the world by 2004

2006 Statements Opposing the Doctrine of Manmade Global Warming
- put on record by - U.S. Senate Committee on Environment & Public Works

2007 The U.S. Senate Report:
- Over 400 Prominent Scientists Dispute Man-Made Global Warming Claims - listed by name in detail

2008 New Oregon Petition Project - online, and still ongoing - signed by over 31,000

Details at: www.ice-age-ahead-iaa.ca/alternate_healing/lovescapenovels/climate_change_opposition.html

The big protests began in 1992 with the Heidelberg Appeal from the University of Heidelberg in Germany. The appeal was signed by 4000 scientists. It was followed up in 1997, in Germany, with the Leipzig Declaration project that addressed the narrow field of actual climate specialists. It was signed by 110 of the specialists.

Later, in 1998 The Oregon Petition Project was staged in the USA. (see details at: http://www.oism.org/pproject/s33p36.htm) The petition project was signed by 17,000 scientists worldwide. The project was organized in opposition of the UN's Kyoto Protocol on Climate Change, and appears to have had an effect. The Kyoto Protocol eventually met with an 85% rejection across the world by 2004.

As this didn't end the global warming politics, numerous open letters, including to governments, were issued in 2006 by prominent scientists opposing the Doctrine of Manmade Global Warming. Then in the next year, in 2007, the high-level U.S. Senate Report was compiled that comprised detailed statements from over 400 prominent scientists, disputing the Man-Made Global Warming claims.

A year later, in 2008, a new Oregon Petition Project was staged online. It received over 31,000 signatures, and is still ongoing.

For details, see: www.ice-age-ahead-iaa.ca/alternate_healing/lovescapenovels/climate_change_opposition.html

For a summary, see: www.ice-age-ahead-iaa.ca/alternate_healing/2011iceage/ice_age_mass-protests.html
Also see the video: "Science Community Opposition" http://www.ice-age-ahead-iaa.ca/19/index.html

Global Warming opposition is actively supporting the Global Warming hoax

War men! War men!

But no fighters for humanity

War men! War men!

But no fighters for humanity

It appears, in retrospect, that the overwhelming consensus in the science community stands in opposition to the manmade global warming doctrine, and continues so, while in real terms the entire Global Warming opposition movement is actively supporting the Global Warming hoax, by supporting its fundamental premise, the invariable Sun.

None of them fight for humanity by focusing on its most crucial issue.

The crucial issue is, the Sun is NOT its own master

The crucial issue is, that the Sun is NOT its own master, as it varies with external conditions, which thereby affect us, including with Ice Ages. This means that the invariable Sun is an illusion that needs to be overcome.

The invariable Sun simply does not exist in the real world. The illusion of its existing is build on the prior illusion, that the Sun is its own master, operating disconnected from the universe.

The Sun is falsely perceived

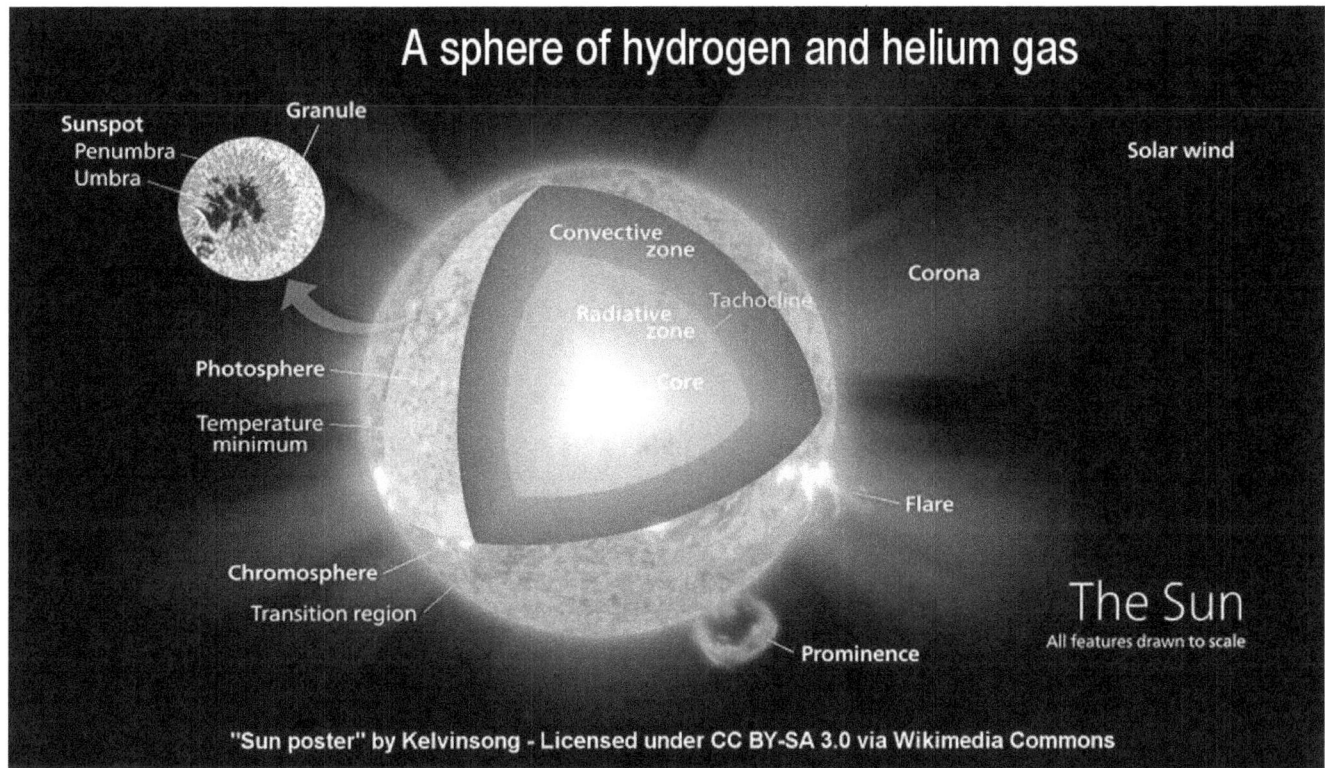

The Sun is falsely perceived as an independent nuclear fusion furnace that 'burns' hydrogen gas into helium gas at a steady, invariable rate. The Manmade Global Warming doctrine is built on the false, steady-Sun, premise in order that all climate changes can be declared to have been manmade, which is politically required as a factor of control.

The opposition movement accepts the invented basic model of the invariable Sun. It does NOT dispute this model. It accepts thereby the very premise that the Manmade Global Warming hoax is built on. The Global Warming Opposition movement only disputes the numerous little details of false assumptions that the Manmade Global Warming doctrine espouses, instead of it disputing the core issue - the false model of the Sun - that the hoax is erected on.

Science Betrayal? (Fighting with Clubs)

Science Betrayal?

(Fighting with Clubs)

Science Betrayal? (Fighting with Clubs)

No one speaks for humanity in this senseless climate war

Francisco de Goya - Fight with Cudgels', c. 1820–1823. Oil mural

On the resulting huge war front of science pitted against science, warring over concepts that are not real, the Ice Age focus remains completely out of sight to the present day, while the warriors remain grounded in mud, locked in war with each other that no one can win, as in Goya's painting.

To win, means to step away from the illusion that locks the opponents into war.

No one speaks for humanity in this senseless climate war, much less fights for humanity - which is in grave danger on the most critical issue of all time in the Ice Age boundary zone.

No one is fighting for scientific truth in this fight - the truth of the now evermore weakening Sun in the plasma powered universe; the type of Sun that we have numerous measurements of to prove it - which now endangers our agriculture, while no Plan-B is presently considered to enable humanity's continued existence.

The Truth has been Late Unfolding

The Truth

---

has been Late Unfolding

The Truth has been Late Unfolding

It wasn't until modern time

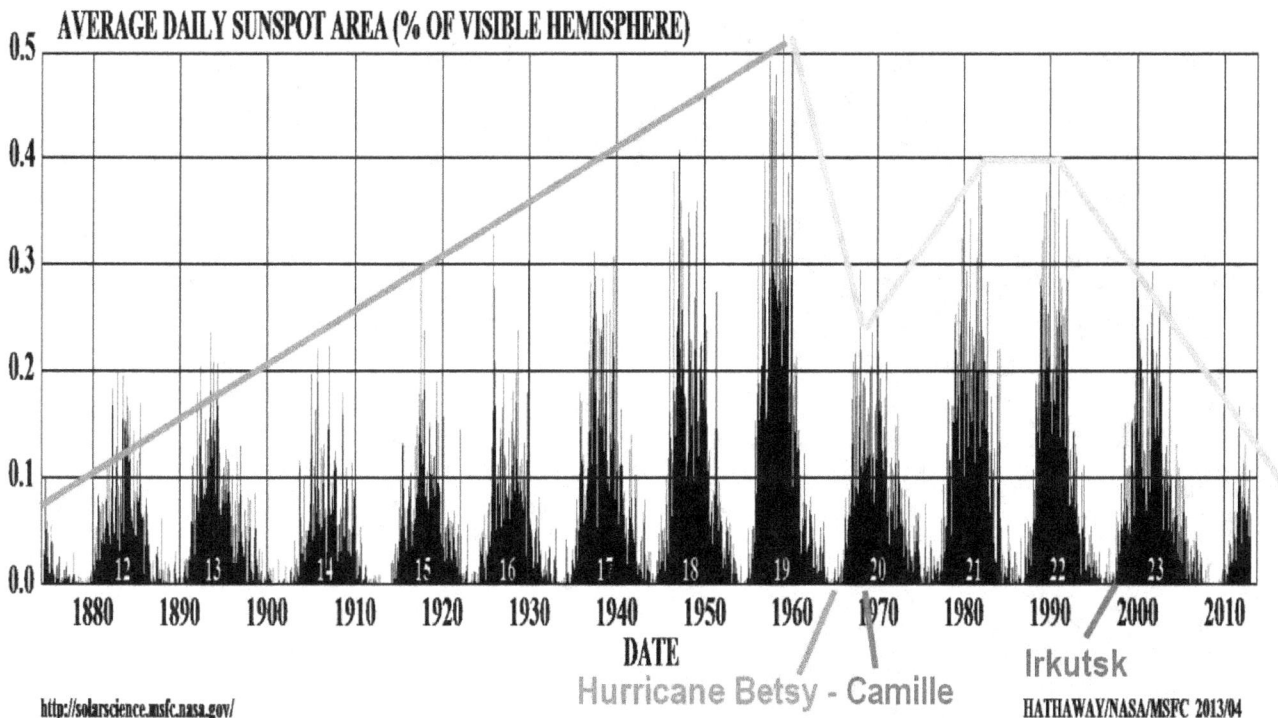

Actually, it wasn't until modern time, and with the aid of space age technology, that a scientific basis for the near Ice Age was gradually being developed. In the 1970s the scientific community only saw the sunspot numbers sharply diminishing, without being able to understand, why? The scientific data for this answer didn't exist at the time.

Some people may have had a hunch that the Sun was the central actor

SUNWARD BOUND

By NATIONAL AERONAUTICS AND SPACE ADMINISTRATION
FOR RELEASE: May 16, 1978 No. 78-H-237, 78-HC-189

Sure, some people in the scientific community may have had a hunch that the Sun was the central actor on the climate scene and the Ice Age. It may have been against this potential background that the Ulysses Mission to orbit the Sun had been vigorously pursued in the 1970s and 1980s.

When Ulysses was finally launched in 1990

Ulysses was launched from the NASA Space Shuttle Discovery on Oct. 6, 1990

NASA

And when Ulysses was finally launched in 1990, after a long struggle against budget cuts, it took another 18 years until its revolutionary measurements had been completed that opened up a new era of looking at the Sun.

In the shadow of Ulysses

NGRIP - The North Greeland Ice Coring Project (1996-2003)

funded by institutions in Germany, Japan, Sweden,
Switzerland, France, Belgium, Iceland and the US.
Primary sponsor is the Danish Research Council.
(see: http://www.gfy.ku.dk/~www-glac/ngrip/index_eng.htm)

In the shadow of Ulysses, also some of the biggest ice coring projects were undertaken with amazing results that did add to our understanding of the solar dynamics.

Likewise did plasma physics experiments add immensely

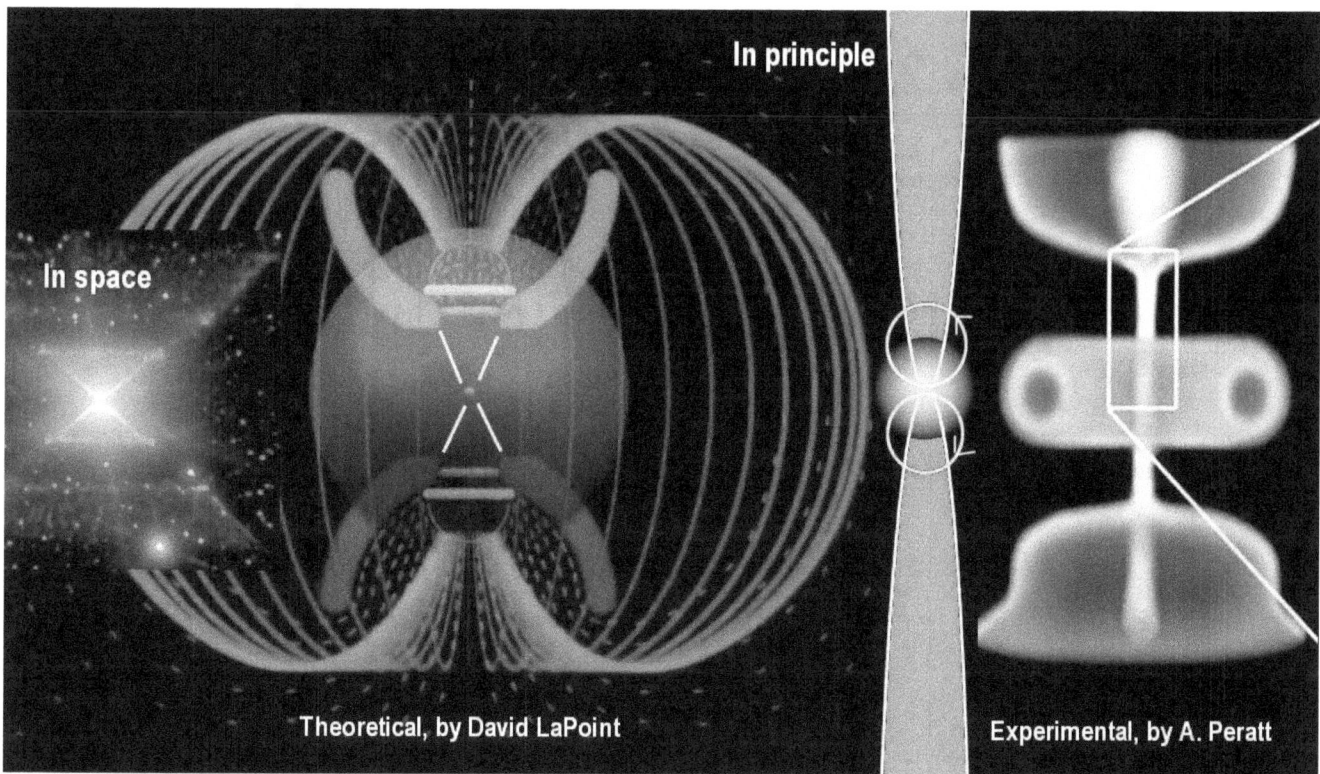

Likewise did some of the largest and smallest plasma physics experiments add immensely in recent years to our understanding of the solar system and the Sun. But in spite of all these huge scientific successes, nothing came out of it that changed the world. The data became slowly buried.

No breakout is yet in sight

No breakout is yet in sight, from the war
over Manmade Global Warming Climate Change

No breakout is yet in sight, from the war over Manmade Global Warming Climate Change

In spite of the immense progress that has been made

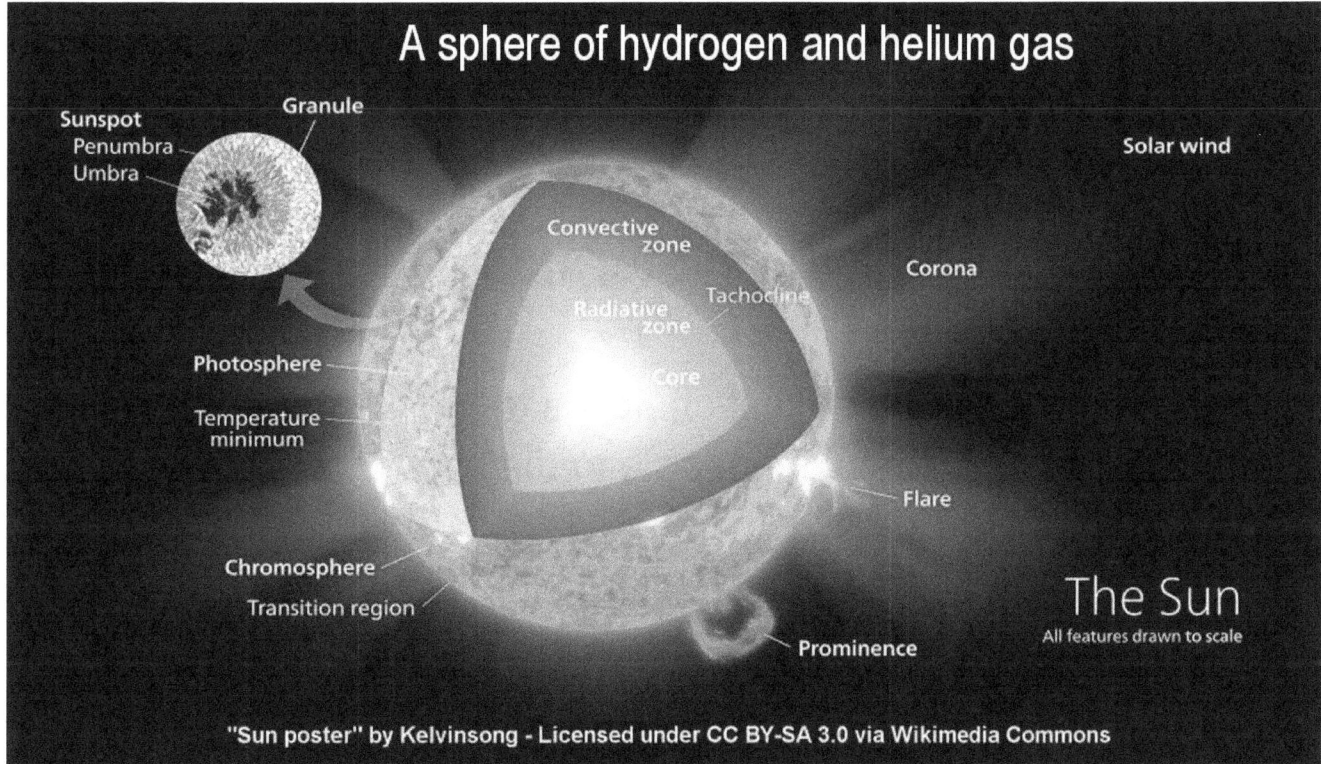

In spite of the immense progress that has been made in plasma astrophysics, both of the global warming combatants, pro and con, remain trapped as before, in this ideological box that few dare to escape from, or even wish to escape from. The trap is the tightly wound knot of the hydrogen-fusion model of the Sun that all bow to, while all evidence disputes it, but which is required politically. It is politically useful to declare all climate changes to be manmade, with scare stories attached which opens the door for control over humanity.

By launching its double trap of manmade global warming and the hydrogen fusion sun model, the imperial oligarchy did indeed come to own the sciences and control the thinking of society and its actions, and in doing so did indeed effectively block humanity's future as every imperial system requires, for which the trap was set up back in in 1975.

Plan-A on steroids:

Plan-A on steroids

The official choice is to force the people to die

Plan-A on steroids: The official choice is to force the people to die

The global warming project continues to operate as a trap

A shocking statement was made by a United Nations official Christiana Figueres at a news conference in Brussels. Figueres admitted that the Global Warming conspiracy set by the U.N.'s Framework Convention on Climate Change, of which she is the executive secretary, has a goal not of environmental activists to save the world from ecological calamity, but to destroy capitalism. She said very casually:

"This is the first time in the history of mankind that we are setting ourselves the task of intentionally, within a defined period of time, to change the economic development model that has been reigning for at least 150 years, since the Industrial Revolution."

She even restated that goal ensuring it was not a mistake: *"This is probably the most difficult task we have ever given ourselves, which is to intentionally transform the economic development model for the first time in human history."*

posted in a blog on Climate, Feb 3, 2017 by the renowned American economist Martin Armstrong.

"Global Warming is about Destroying Capitalism"

https://www.armstrongeconomics.com/
world-news/climate/
global-warming-is-about-destroying-capitalism/

So it is, that to the present day, as from the very beginning of the global warming project, the project continues to operate as a trap, devoid of actual science and truth, but strong on control and genocide. The objective remains, and is openly stated, to promote the genocide by shutting down human development, as the objective had been in the beginning. As of today, the objective is succeeding.

The modern revelation

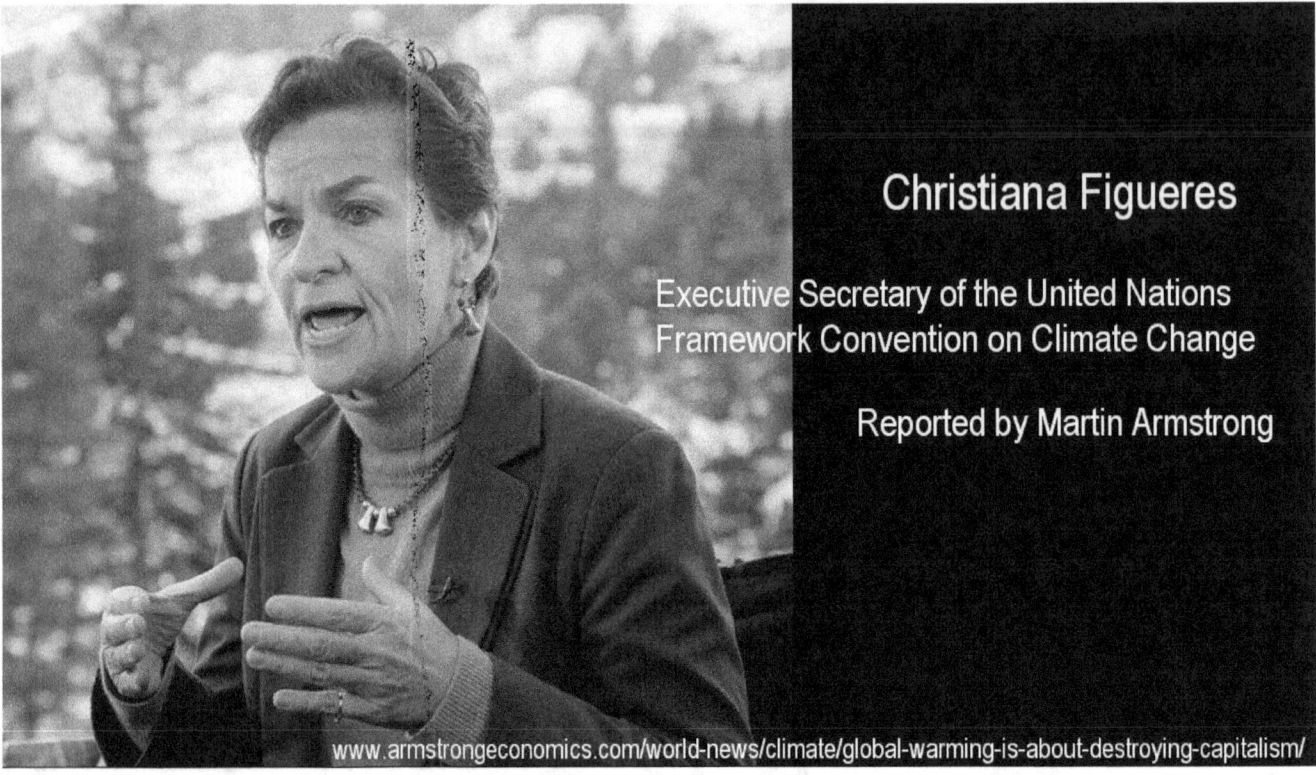

The modern revelation, in this case, was announced from the very top by the Executive Secretary of the United Nations Framework Convention on Climate Change.

She is denouncing the model that has created the USA as a prosperous economic power

Mr. Armstrong quotes Christiana Figueres
Executive Secretary of the United Nations Framework Convention on Climate Change

"This is the first time in the history of mankind that we are setting ourselves the task of intentionally, within a defined period of time, to change the economic development model that has been reigning for at least 150 years, since the Industrial Revolution."

The high-level U.N. secretary is quoted, saying about the modern objective, "This is the first time in the history of mankind that we are setting ourselves the task of intentionally, within a defined period of time, to change the economic development model that has been reigning for the last 150 years, since the Industrial Revolution."

She is denouncing the model that has created the USA as a prosperous economic power, and Europe likewise in many respects, which had build a foundation for humanity to become what it is, and to have a future.

Tragically, the U.N. project is succeeding

Emblem of the United Nations

The power-house for our modern living, the industrial revolution, is now slated to be torn down, intentionally, worldwide, while the opposite should be on the agenda. Tragically, the U.N. project is succeeding wonderfully well. The western economies are de-industrializing; poverty is rising; infrastructure are collapsing; crime, dope, and terror are increasing; and science is so tightly tied into knots, that the real astrophysical climate dynamics are completely off the table as if they didn't exist.

In this mentally imprisoned, small-minded ,top-level Malthusian world

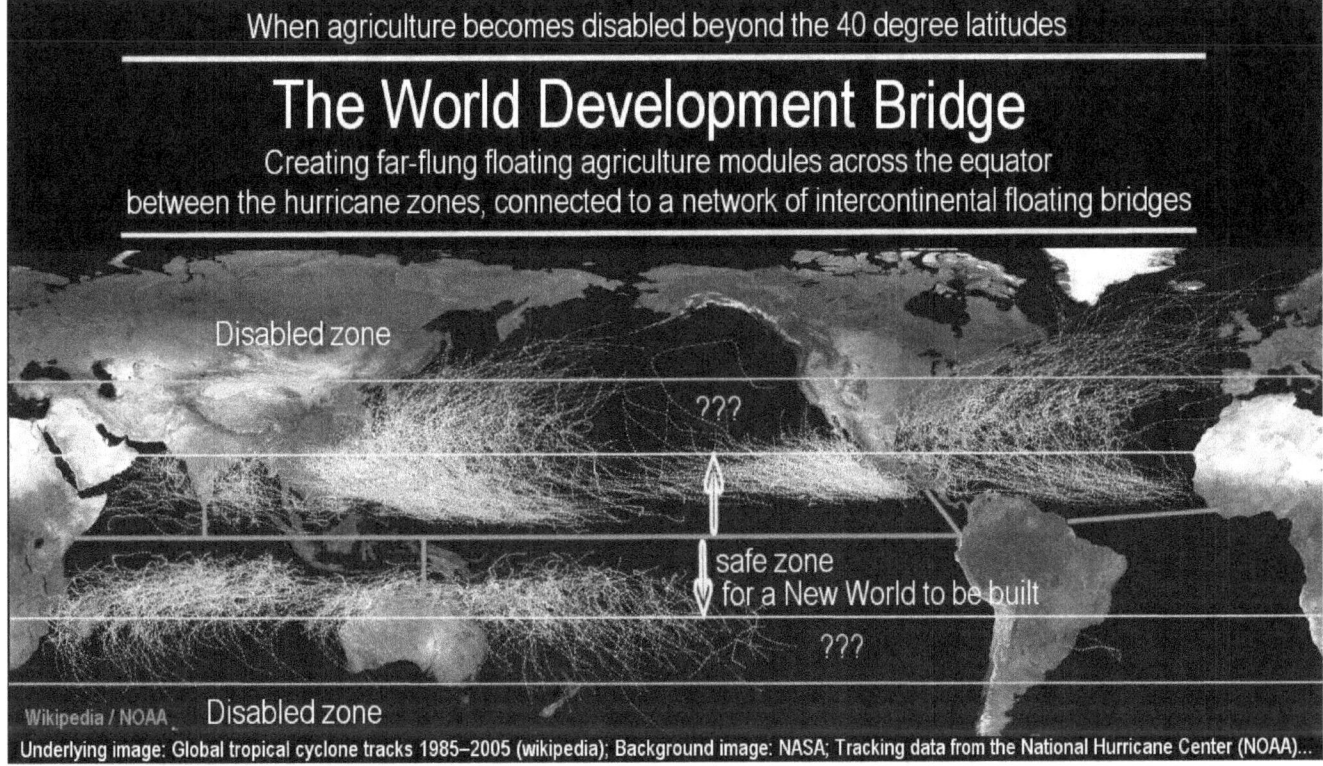

No one raises an eyebrow anymore in this mentally imprisoned, small-minded ,top-level Malthusian world, when in the real world human existence as a whole is in danger of the Ice Age phase shift and its consequences. Even while time is running out, no one talks about a Plan-B World-Bridge development to assure humanity a future being built with a super-industrial revolution employing new energy systems, new materials, new agricultures afloat on the Equatorial seas, complete with 6,000 new cities attached rent free, and so on - with which to assure the continued existence of humanity on the type of Ice Planet that the Earth will largely become when the boundary zone ends in the 2050s or sooner, and the Ice Age begins that no one can yet imagine much less survive without a Plan-B.

Neither are any eyebrows raised

Near the United Nations
General Assembly Hall

Neither are any eyebrows raised in the halls of the U.N. or in the halls of governments when the existence of entire nations is in the process of ending - potentially within the next 10 years, as is presently in progress - when agricultures will collapse, and housing and transportation will begin to collapse likewise, without a Plan-B implemented to replace them.

Stepping Away from the Pantheon of Mediocrity

Stepping Away from the Pantheon

of Mediocrity

Stepping Away from the Pantheon of Mediocrity

A debating center for a pantheon of mediocrity

United Nations
General Assembly
Hall

What is not talked about in the gilded halls yet, is the truth that we would hear from the U.N. loud and clear if the U.N. lived up to its design, which it may do some day.

Instead, presently, the Ice Age challenge together with its imperative, remains hidden there, and this as effectively as when it had been buried in the 1970s when the manmade global warming was invented and made a U.N. project.

This sad situation will likely not change for as long as the U.N. operates as merely a debating center for a pantheon of mediocrity that is easily controlled by imperial interests.

Plan-B started unilaterally, as a pioneering venture

Belt and Road Forum 2017 gala

It may well be, therefore, that the Plan-B infrastructures will be started on a different platform, unilaterally, as a pioneering venture by a country like China that is coming alive with creative optimism, honest science, and is by its Confucian cultural background committed to the advance of the general welfare of its people and beyond that of the world.

We see a bit of that already shining through

https://www.youtube.com/watch?v=zJHFM7xsPgM
image from "Belt and Road" closing Gala, televised by "New China"

We see a bit of that already shining through on the world scene with China's "Belt and Road" infrastructure development project making huge strides forward, that over a hundred countries already participate with, with evermore joining.

Once the dynamism spills over onto the Plan-B World Bridge development

Chinese-built Mombasa-Nairobi Standard Gauge Railway, Kenya

Once the already developing dynamism spills over onto the Plan-B World Bridge development project for the security and expansion of food production and free living, millions will line up wanting to participate. Some may even drag their countries behind them, especially when their territories are starting to become disabled, as may soon be the case in Russia, Europe, Canada, and so on.

http://www.xinhuanet.com/english/2017-12/28/c_136857657.htm - Chinese Rail in Africa

Part 9 -
Spontaneous, Unbounded, Cultural and Industrial Revolution

Part 9

Plan-B Unfolding

Spontaneous, Unbounded, Cultural and Industrial Revolution

Part 9 - Plan-B Unfolding: Spontaneous, Unbounded, Cultural and Industrial Revolution

Great efficient principles tend to inspire the needed progress

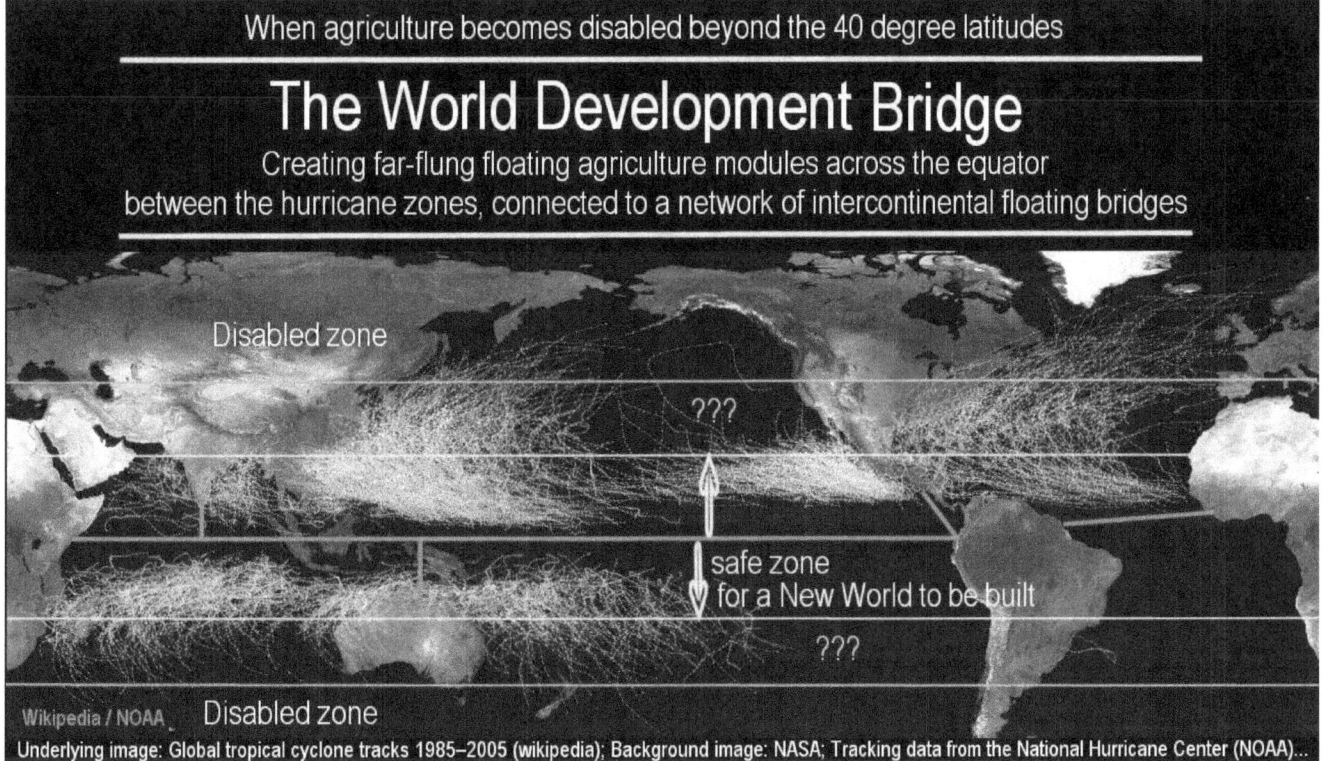

Great efficient principles, founded on the tall platform of truth, tend to become their own imperative and inspire the needed progress.

Thus the question of whether humanity will survive through the boundary zone into the near Ice Age, is not an astrophysical question, nor even a scientific question, but is a question of humanity's self-perception and self-development as human beings.

A question of how much value we attribute to our existence

*The Greatest Miracle on Earth
a human being
born in the image of God
reaching for spiritual light
to understand the universe
and itself*

It becomes a question of how much value we attribute individually to our existence, to our being alive on this planet, and to being a part of this world as creators and producers and writers and artists and farmers. If we find no value in human existence, or too little to bother, then Plan-A will succeed by default and humanity dies.

If, inversely, we find great value in us all as the greatest species of life on the Earth, then a decisive Plan-B will become implemented without fail, with which a New Age for humanity will begin that is not a default result, but is a created result winged with unbounded aspirations.

My perception is that a grand Plan-B will become implemented

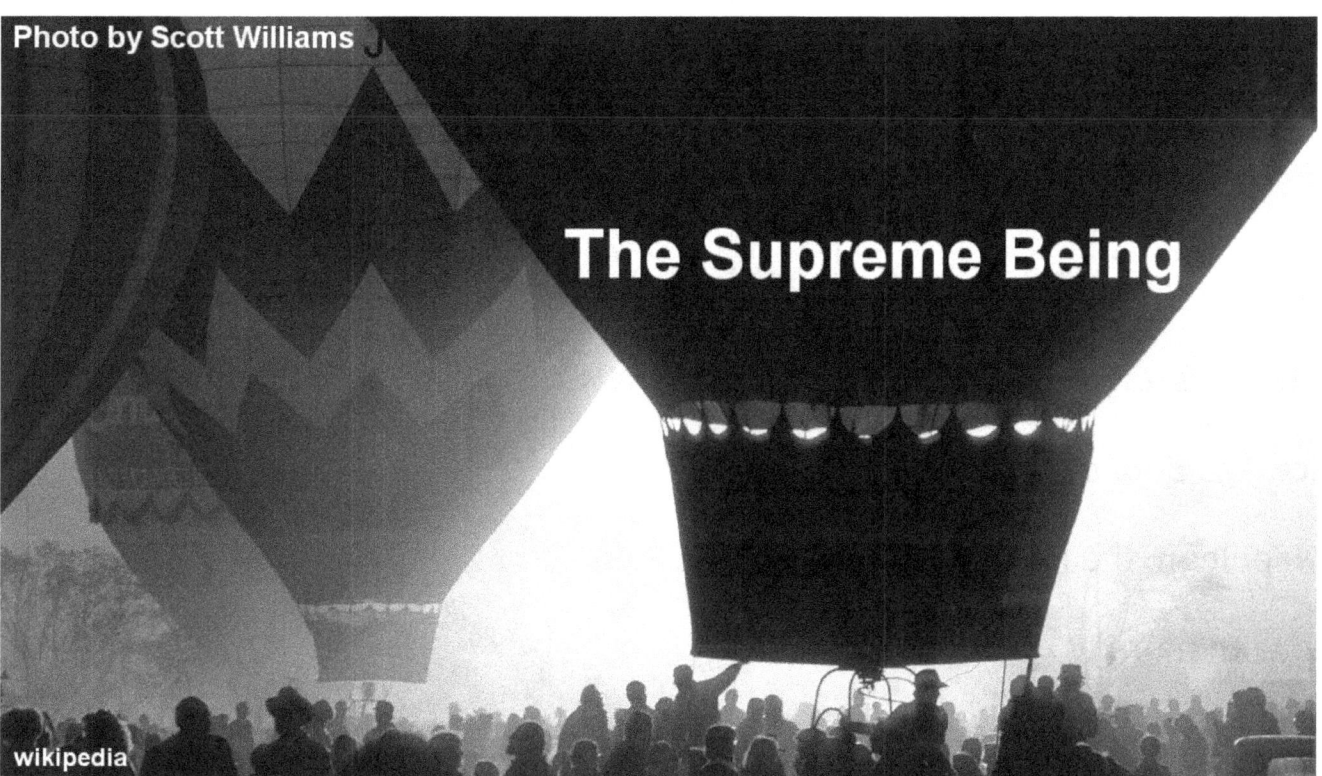

My perception is that a grand Plan-B will become implemented one way or another, because the spirit of reaching beyond the default level to difficult challenges and near impossible adventures, has always been the hallmark of our human society. We are the giant in this field, and we are awaking. Once we are awake, it will be but a small thing for this giant that we are, to move with the movements of the universe, and move miles ahead of it and above it, which we have the capacity to do on the wings of science unfolding.

Similar Books in print,
by Rolf A. F. Witzsche

Grand Solar Minimum Becomes the Ice Age

Ice Age Uncertainty

The Amazing Sun Literacy

Ice Age Boundary Zone

Napoleon, the Ice Age, and the Empire Sun

... with more to come

For details, see (books - Ice Age Science Illustrated):

http://www.ice-age-ahead-iaa.ca/57/index.html